THE HUMAN MODEL:
PRIMATE PERSPECTIVES

THE HUMAN MODEL:
PRIMATE PERSPECTIVES

HARRY F. HARLOW and CLARA MEARS
University of Wisconsin and University of Arizona

1979

V. H. WINSTON & SONS
Washington, D.C.

A HALSTED PRESS BOOK

JOHN WILEY & SONS
New York Toronto London Sydney

V. H. Winston & Sons, a Division of Scripta Technica, Inc.,
Publishers
1511 K Street, N.W., Washington, D.C. 20005

Distributed solely by Halsted Press, a Division of John Wiley
& Sons, Inc.

Library of Congress Cataloging in Publication Data

Harlow, Harry Frederick, 1905—
  The human model.

  Bibliography: p.
  Includes index.
  1. Psychology, Comparative. 2. Primates—
Behavior. I. Mears, Clara, joint author.
II. Title. [DNLM: 1. Ethology—Collected works.
2. Models, Psychological—Collected works.
3. Primates—Collected works. QL737.P9 H286h]
BF671.H37   156   78-27597
ISBN 0-470-26642-2

CONTENTS

FOREWORD vii

PROLOGUE–The Human Model 1

PART I–The Legend of Learning

Chapter 1. The Development of Primate Testing 13

Chapter 2. The Development of Learning 27

Chapter 3. The Evolution of Learning 43

Chapter 4. Learning to Learn 61

PART II–The Meaning of Motives

Chapter 5. The Matrix of Motives 77

Chapter 6. Monkeys, Mice, Men, and Motives 89

Chapter 7. The Nature of Love   101

Chapter 8. The Nature of Love Simplified 127

Chapter 9. The Power and Passion of Play 141

PART III–The Loves of Life

Chapter 10. The Linkage of Loves 161

Chapter 11. The Basic Tryad 173

Chapter 12. Peer Persuasions 189

Chapter 13. Heterosexual Love 201

51412

PART IV—The Price of Pathology
Chapter 14. Pathological Perspectives   217
Chapter 15. Broken Bonds   225
Chapter 16. The Hell of Loneliness   241
Chapter 17. Love Restored   257

EPILOGUE—From Thought to Therapy   273
BIBLIOGRAPHY   297
INDEX   309

FOREWORD

The birth of this book was probably less painful than the birth of many books with a scientific inheritance. I am sure that its conception, however, was far more protracted than are many births. This book was not at first but a gleam in its father's eye. I am father to the thought but not the book. This book was first a persistent cortical and then vocal insistence on the part of a former graduate student, Stephen Bernstein, who some years ago nurtured me through his Ph.D.

Bernstein was convinced that a certain select number of my lifetime professional publications should be preserved and presented for perusal under one coat. The plan was to select the desired publications and reproduce them in full and in the original form. To this effect, the selection of thirty publications was made. Upon thorough review of these former articles, Bernstein, with the collusion of my wife, Clara, began questioning me concerning experimental findings of the subjects covered not only in these articles but antecedent and subsequent to them. Tape recorders were our constant companion on travels over the United States.

The exigencies of length required the reduction of the number of originally selected articles to be reproduced in entirety. Others of the selection have been excerpted and the permission of the copywriters gratefully acknowledged. Through no fault of our own, but through the time, efforts, and the prodding of our tormentors, the balance of the chapters were written anew to include the findings of many more than the original articles, all of which are indicated in the bibliography by asterisks.

Dr. Stephen Bernstein, now a professor at the University of Colorado, made an additional unique contribution. He persistently managed to direct both authors to the proper medical sources to maintain good health until the completion of the book.

Evident throughout the book's figures is the wit and perspicacity of Robert Dodsworth, for many years the Director of the Photographic Laboratory at the University of Wisconsin Psychology Primate Laboratory.

Special acknowledgement is tendered to both Mrs. Helen LeRoy and to Dr. Steve Suomi, pillars of the Primate Laboratory, and to the successive years of graduate students whose contributions have been inestimable.

The following acknowledgements are presented gratefully for material reproduced in full or substantial entirety: Harlow, H. F. The evolution of learning. In A. Roe & G. G. Simpson (Eds.), *Behavior and Evolution*. New Haven. *Copyright* (1958) *by Yale University Press. Reprinted by permission* in Chapter 3. Harlow, H. F. Mice, monkeys, men, and motives. *Psychological Review*, **60**, 23-32. *Copyright* (1953) *by the American Psychological Association. Reprinted by permission* in Chapter 6. Harlow, H. F. The nature of love. *American Psychologist*, **13**, 673-685. *Copyright* (1958) *by the American Psychological Association. Reprinted by permission* in Chapter 7. Harlow, H. F., & Suomi, S. J. Nature of love—simplified. *American Psychologist*, **25**, 161-168. *Copyright* (1970) *by the American Psychological Association. Reprinted by permission* in Chapter 8. Harlow, H. F., Harlow, M. K., & Suomi, S. J. From thought to therapy: Lessons from a primate laboratory. *American Scientist*, **59**, 538-549. *Copyright* (1971) *by the Society of the Sigma Xi. Reprinted by permission* in the Epilogue.

Harry F. Harlow
October 1978

PROLOGUE

THE HUMAN MODEL

Throughout my entire academic life studying rhesus monkeys I had never suffered for research ideas, since I simply stole the research ideas from human studies or from human problems. I always believed that I should never do anything with monkeys that would not have significance with man, as is brought out in the Epilogue, "From Thought to Therapy." I firmly believe that one should never study problems in monkeys that cannot be solved in man. What direction my research might have taken had this not been true, I have no idea.

Research on subhuman animals may be conducted for multiple and different reasons. It may be a search to sort the similarities, dissimilarities, and contiguities as the primate orders ascend or to compare the behavior of animals belonging to entirely different orders. Research with subhuman animals may attempt to penetrate the principles of primary areas of behavior, such as motivation, learning, emotions, psychopathology, or treatment of disorders.

A question commonly asked is, "Does primate research generalize to human beings?" This question presents an unsolvable problem which makes

1

Fig. 1. Human model of self-motion play.

the question academic. At the Wisconsin Primate Laboratory we have approached the question in reverse. Because of the basic similarity between monkeys and men, human models may be used to plot and to plan monkey researches, aside from the human being who creates the research. There is far more reason to use human data to plan and interpret monkey researches than to use monkey data to plan human studies. Long research experience has taught us that adequate human data generalize with pride and prescience to monkey data. If human data generalize to monkeys, we are not particularly concerned whether or not monkey material generalizes to man—of course the data do.

For example, the standard human learning model is that of maturation of the intelligence of the IQ. Monkeys are obviously not as intelligent as human beings, but they have their IQs too. The data that make possible the human IQ scores, based on the maturation of learning ability, fit very well with the monkey data. Monkey intelligence matures in the same way as human intelligence except in two respects. Monkey intelligence matures more rapidly

Fig. 2. Rhesus monkey self-motion play.

and, tragically, stops much sooner. The one-year-old monkey is about twice as intelligent as the one-year-old human being, but eventually almost all human beings attain or even surpass the monkey's intellectual capabilities. The tragedy is probably greater for our research than for the monkey. We are far less concerned about the relative intelligence of monkey and man than the fact that the nature of the growth of intelligence is comparable in monkeys and men. This is an example of taking a human model and finding out that it works perfectly in the monkey, making proper allowances for growth rates, brain size, and cortical complexity.

There are many reasons for studying the behavior of monkeys. When monkeys, through the process of evolution, came to mimic man in body and in brain, the best research animal available for studying man became the monkey, and this is an accepted fact for both medical and psychological research.

There are a number of enormous advantages in the use of monkeys rather than man for research purposes. In spite or because of inflation, and even with the necessity of raising one's own monkeys, monkeys will always be cheaper subjects than men. One may profitably study monkeys to obtain basic and beautiful information about people. Monkeys are not little men, but they sought to achieve this admirable status for about 40,000,000 years. Unless man capriciously destroys the environment and most of its inhabitants, the monkeys will have ample time to evolve into little men. Actually this is a very minuscule motive. If the monkeys do evolve into little men, they will have a long tale to tell and little tail to use.

The hours that monkeys function in fruitful fashion are determined by the experimenter, whereas human subjects' working hours are determined by labor unions, often belabored unions. Strikes are nonexistent in the simian economic theory, and hunger strikes especially were not part of their Indian culture. The life span, particularly the test-effective life span, is abbreviated in monkeys compared to man. Longitudinal studies may be done within the working life of one experimenter. Tests may be made on many generations of monkeys. Monkeys are manageable enough in the laboratory that they can be fed and they can be bred. Being fed is very good for the monkeys, but has limited scientific satisfaction. Breedability offers enormous substance and sustenance, for controlled breeding forms and corner and keystones for scientific success in a host and wealth of scientific endeavors.

Between 1955 and 1960 we created our own breeding colony at the University of Wisconsin and thereafter had age-dated pregnancies which produced age-dated babies. Age-dated babies make possible age-controlled subjects for scientific studies. Fortunately, human behavior does not change so fast that we produce age-dated results. Many behaviors, even diverse behaviors ranging from learning to sex, including sexual learning, mature in monkeys at 4 to 5 times the human rate. In terms of anatomical age, as assessed by anatomical criteria, monkeys at birth are equivalent to human infants a year of age. Knowing monkeys and human chronological age, one can trace and compare the maturational development of endless, interesting, buoyant behaviors including intelligence, social play, emotionality, and all of the forms of human love, mother love, baby love for the mother and peer or agemate love, to say nothing of parental love and mature love between the two sexes.

When we are comparing the behavior of various animals, particularly various primates, the point of proper theoretical regard is the generalization from man to monkeys. However, monkey data may still have its own and proper place in the sun. Human researches, whether planned or accidentally propagated, are beset by problems of constraints and controls. People are restless under regimes of regulation, they reserve cage confinement for criminals, their early life histories are often directionless or diverse, and they righteously and relentlessly resist abnormal environments such as isolation or abnormalities such as deliberately induced depression or schizophrenia.

Macaque monkey research, conducted in a humane manner, may be properly used to extend human research which must of necessity be limited by normal and natural restraints.

One can separately test a multitude of variables in rearing conditions when doing monkey investigations. One can raise monkeys with their natural mothers, or with adopted mothers, or on completely artificial surrogate mothers. One can raise baby monkeys with loving mothers, with inattentive, non-cuddling mothers, or even neglectful mothers and discover the varying effects on the social development of the child. If abberrations in the mother's own background should produce a cruel or an abusive mother, there are protective, knowledgeable attendants to protect the infant instead of being, like the human child, left to the unmerciful, mangling or fatal treatment of the human abusive mother. All infants have the right to be raised in the best possible way. With monkey infants we can find out what is the best possible way.

Common sense says that we should raise all children in an optimal environment. But what is the optimal environment—or the so-called normal environment? The normal environment for a mole is a hole, for the snake, a pit, and for the wolf, the wild. The normal environment for any animal is the living space where he breeds and is bred, where he feeds and is fed, and where he successfully goes through and grows with all the love systems. Many behavioral scientists, particularly ethologists, believe or believed that the only normal environment was in the animal's natural habitat, in its indigenous feral home. The ethologists who have held this position have never lived in a feral environment and, indeed, if they had they would have been sacrificed. Civilized man's so-called "normal environment" is the ever-changing experimental cage where he has lived for 2,000,000 years as the survival of the fittest and the fleetest determined. Had man lived forever in a feral environment, he would never have invented Kleenex, comfort, and the pencil eraser. Indeed, countless millennia ago he himself would have been thoroughly erased. Man was the first known animal to successfully leave the feral environment on his own initiative and to modify the environment to his needs instead of modifying himself to the environmental demands.

Why shouldn't this be true? Our own data show without question that the monkeys we raise in the laboratory are healthier, less disease-prone, more fearless, brighter, and presumably happier than any monkeys ever raised in a wild or feral state. These monkeys may not be in the "natural state" visualized by the ethologists, but they are in their element, in surroundings and situations suited to them. Human laboratory mothers or fathers are brighter and better trained than monkey mothers. Human caretakers are motivated to improve their methods, and unfortunately monkey mothers and fathers cannot read. Trained human beings know more about infant monkey diseases than the most highly educated monkey mothers. They have recourse to vitamins, modern medicines, the secrets of sanitation, and veterinarians. Even in modern times it was taken as fact that one child in seven would not survive the first year of life. Probably only half of the monkey infants born in the forests of India ever have a first birthday. In our laboratory, about 95% of our laboratory born monkeys lived through puberty.

It is commendable to study subhuman animals in the feral environment, but it is better to do so in a laboratory where there are controls, as long as you do not destroy the capability for social interaction. The challenging problem is to produce laboratory environments in which the feral animal transcends its feral capabilities. What man did for man he should be able to do for monkeys or for any other laboratory animal. As a matter of fact, we often and deeply hope that man may in many instances do better for the monkey than man does for man.

The first major researches at the Wisconsin Primate Laboratory related to cortical localization of function. Through lesions in varied locations of the cortex, different intellectual functions could be related or localized to appropriate cortical areas. In the 1920s Lashley had postulated equipotentiality of all areas of the cortex for all intellectual functions. A decade or so later, Jacobsen conducted his classical studies on localization of function in the prefrontal lobe of the cortex. He showed that destruction of this area was followed by loss of immediate memory as measured by delayed response tests. These data were consistent with the loss of recent memory in senile patients whose autopsies revealed degeneration in the prefrontal lobe.

Since we had age-dated monkeys, comparison of intellectual loss through identical lesions at different ages could be assessed. In one group an operation on the frontal lobe was performed when the monkeys were just under 1 year of age. In the second group the same procedure was followed when the monkeys reached 4 years of age and had attained mental maturity. At 4 years of age, the frontal lobectomy totally destroyed delayed response capability. After the same operation at just under 1 year of age, the monkeys exhibited near-normal performance on all tests, including the delayed response tests.

In other words, identical anatomical lesions produce entirely different syndromes when the lesions are made at disparate ages of 1 and 4 years, respectively about 4 and 14 years of age in human life. Thus, brain injuries from falls or as a result of auto accidents can produce far different apparent behavior or brain loss at the various developmental ages. If you are going to injure the brain, do it young when there is a chance of much greater recovery and much less original loss. Neonatal brains are apparently more plastic, and intellectual abilities which are highly localized at intellectual maturity are diffusely represented in the entire *neocortex* during early infancy.

Karl Pribram (1955) at the Yerkes Laboratories and Harlow (Harlow,

Schiltz, Blomquist, & Thompson, 1970) and associates at the Wisconsin Primate Laboratory identified a second brain or cortical syndrome during the period from 1955 to 1970. Large cortical lesions in the superior and posterior temporal lobe gave rise to no delayed reaction impairment but to very significant loss on tests of discrimination and discrimination learning set. Similar effects have been reported for human patients, and this phenomenon has been described by Teuber (1955) as the double dissociation of symptoms.

The human clinical data on the symptoms associated with the two differential brain lesions described is far less precise and definitive than the monkey experimental data. There do not, however, appear to be any significant discrepancies. This is a common comparison between human and monkey studies—there are seldom, if ever, disagreements, but the better controlled subhuman studies are more exact and decisive.

There are marked differences between monkeys and men when brain lesions develop or are incurred in the preoccipital or posterior parietal areas. Since these differences relate to language deficits, the differences are not surprising. Man is the only language facile animal. Lesions involving the angular and supramarginal gyrus on the left side of the brain produce no loss in monkeys, but severe language comprehension losses in man. Head (1926) terms these language deficits syntactical and semantic aphasia. Lesions in the preoccipital areas probably give rise to a third or fourth basic syndrome of intellectual loss, which is under investigation by Blomquist at the present time.

Injuries or lesions causing impairment of function or destruction of cortical areas may have surprisingly specific effects. Combined frontal and temporal lesions impair the ability of the monkey to achieve learning sets when the stimuli differ only in form. These same lesions have little effect if the stimuli differ only in their color. Color discriminations are simpler than form discriminations. Since the learning set is one of the most useful of learning tools for both monkey and man, any disruption can cause widespread confusion. One of the more complex types of learning problems involves acquisition of a learning set for oddity problems. Disruption of the formation of these complex learning sets requires impairment or destruction of the combined frontal and posterior association areas.

The motivation of man and monkey, the drives and desires whose external incentives largely determine behavior, encompasses much more than the satisfaction of the internal drives of hunger and thirst or the avoidance of pain and fear. Human beings, for instance, are motivated by curiosity, the search for the new, the different, and the interesting. After working with monkeys we more and more appreciated the fact that monkeys also were not primarily energized by their digestive systems. You cannot starve a monkey and then tease him with food to tempt him to solve problems. Indeed, the best method is to feed him before he starts to think about problems. Monkey motivation is, like that of the human being, challenged by curiosity and exploration in their own right and sparked by the stimulation of objects in the outside world.

The so-called primary drives of hunger, fear, rage, and pain are actually socially disruptive, not the proper prerogative upon which to form the foundation of behavior of social animals such as men and monkeys. To determine the social motives of monkeys, you review the social mechanisms of man and then the monkey, and you find enormous, basic similarities. The most fundamental social motives are the various forms of love or affection with many comparable components even though man's motives may be more subtle and more persistent.

The interlocking affectional bonds of both of these primates develop through the same stages, starting with the two-way bonds between the child and the mother, the father, or other consistent caretaker, progressing to an affectional system constantly being better understood and appreciated, the agemate or peer system. Until recently the preeminence of the peer system has been little studied in man and for many years has been righteously relegated to the realm of childhood. Upon adulthood, of course, one matured and put away childhood playthings such as toys, and playfellows, such as agemates.

The agemate affectional system was, indeed, stressed earlier and more appropriately in monkey studies than among human beings. Agemate affection is the best socializing system since it involves learning to feel love for multiple members of one's social group, irrespective of sex and even irrespective of age. In spite of its faults and fallacies, affectional development can be based on human data and then effectively translated into well-planned and controlled monkey experiments. Monkey research has both advantages and limitations, but the problem should be supplied by people and resolved by monkeys whose behavior can be better analyzed and comprehended with greater clarity.

The adequacy of mature love between opposite sexes depends upon the nature of each and all of the preceding loves of the individual's lifetime. Emotional, personal congeniality as well as the probability of physical felicity and fitness are each affected by both the presence and the prescience of previous affectional relationships.

Primate psychopathology has a community of characteristics between various and divergent species, although one should not assume that all human psychopathology is mirrored in monkeys. Much as they may try to ape them at times, men are still not monkeys. Analogies are, however, most apt to be discovered in infant behavior before diversification is complicated by either maturation or learning, both cultural and individual. At Wisconsin we initiated our research of monkey psychopathology by studying childhood or anaclitic depression and by successfully attempting to reproduce the syndrome in monkeys. In human behavior the anaclitic depression has been thought of as the depression caused by the separation of the mother and her infants, because that is the most frequently seen.

With the great advantage of subhuman research we were able to experiment and find that this depression could also be caused by the separation of agemate friends from each other over periods of time. Such an attempt with human children would be beset by countless difficulties, entirely aside from the fact that families would never let you intentionally create the separation.

One of the fascinating features of primate research is that you never can predict how far or in what direction it will go. Our cortical localization led to learning research. The learning research led to the creation of an artificial mother called the surrogate terry cloth mother. Of course, when we created a mother whom we could control, she led us to research on infant-mother love, and on and on went the wending of the way of the laboratory studies.

Considering that the IQ found a counterpart in the learning of the monkey, we were not surprised to find that discrete parts of the monkey brain are miniatures of man's brain.

The human child goes the monkey just one better in the realm of curiosity, and the difference is at times to the advantage of the monkey mother and father. Whereas the human child will reiterate, "Why, why, why—and what is that?", the infant monkey just proceeds to find out.

The monkey has taught us many maneuvers of play we were not smart enough to discover from the play of the human child, both about the rules of play and the social roles learned in play, but they are all present in both the human and the monkey primate.

Because human beings are so much more complex than the monkey, they have developed, unfortunately, much more complicated byproducts of love and aggression, such as divorce, unwanted children, and atomic wars.

However, we have held to our position that, if a problem is not important enough to man, it is not worth attempting to solve with monkeys.

PART I

THE LEGEND OF LEARNING

CHAPTER I

THE DEVELOPMENT OF
PRIMATE TESTING

The study of learning in the rhesus monkey per se can be an absorbing subject but the ideal, ultimate objective of the psychology of learning is the discovery of the basic laws governing the acquisition, retention, and utilization of learning in the human being. Even if man could conceivably be maintained in controlled laboratory conditions for extended periods of time, an unhappy subject would be an unsatisfactory subject.

Unable to use the preferred subject, man, psychologists have turned to animals other than man in search for generalizations which might be applicable to human learning. The choice of closely related species possessing similar sensory, motor, and homeostatic systems presents the possibility of the greatest adaptability and generalization of results. Such species would be the anthropoid apes and monkeys which belong to man's own order, the primates, and are our closest phyletic relatives. The laws governing the behavior of certain other species might actually differ in such a manner as to obscure and impede the understanding of human learning.

Not only are the appropriate and effective experimental subjects vital to the study, but there must be tests adequate to test learning from the simple

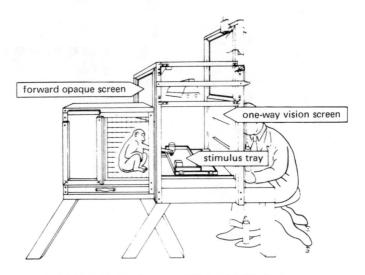

WISCONSIN GENERAL TEST APPARATUS

Showing: Stimulus tray

One-way vision screen in lowered position

Forward opaque screen in raised position

Fig. 3. Wisconsin General Test Apparatus (WGTA).

to the complex and over a developmental time span from infancy to maturity. Prior to the 1930s such tests did not exist for either the right or the wrong subhuman subjects. The breakthrough for a simple reliable learning technique was achieved by Klüver (1933) at Chicago with the creation of the Formboard, and Klüver generously made all information about it available to us.

Ever since the writings of Aristotle, the phenomenon of learning has been related to the phenomenon of association. In writing about association 2500 years ago, Aristotle stressed the two principles, contiguity in terms of time and contiguity in terms of space. In the design of learning apparatus, these principles should always be given full attention, even though higher animals have developed methods making effective associations possible after a considerable temporal delay and with spatial separation between stimuli and responses. Klüver's Formboard (1935) utilizes to the fullest degree Aristotle's two laws of contiguity. The Formboard consists of a flat tray with test objects covering foodwells in which can be placed preferred food rewards.

There is almost a perfect reward for learning, with beautiful contiguity between the stimulus object, the response, and the reward. In 2500 years psychologists had plenty of time to forget these laws. In the original Yerkes apparatus, for instance, the animal ran to the stimulus and then had to turn about face and run in the opposite direction for his reward. In many prior situations, experimenters had not only forgotten the laws of contiguity but had violated them.

We incorporated Klüver's Formboard into the design of the Wisconsin General Testing Apparatus (WGTA) (see Fig. 3). With the discrimination problems Klüver used two foodwells and two objects. The Formboard enabled us, however, to efficiently study multiple kinds of discrimination learning in primates and to devise hierarchal complexities of learning tests. We varied between the use of from two to five foodwells, each covered with test objects. The purpose of this chapter is to present a survey of the problems and the tests for these problems that can be solved by the rhesus monkey and measure learning in subhuman forms and in man. .These problems have ranged in difficulty from the very simple to the very complex as judged by the species of animals which could or could not solve them, and as also judged on logical grounds. The logical ground criterion was always achieved by man since the search for logical language in the fishworm is flimsy and fatuous.

One of the simplest forms of associative learning is the conditioned response as measured by conditioned shock avoidance, conditioned avoidance responses, and conditioned eyelid responses. Even at the level of the learned iridic response, man, monkeys, and lower animals see eye to eye. Although conditioning has had limited study in the subhuman primates, several investigators indicate that conditioned responses may be quickly established with these animals. This question is not of great concern to us since the special value of primates as learning subjects lies in the complexity and range of problems they can master rather than in the rapidity of habit formation.

We have demonstrated in monkeys in five or fewer trials ready reversal of conditioned stimulus-unconditioned stimulus relationships, using fear or other aversive unconditioned stimuli. Harris (1943) reported almost equally efficient conditioning with fewer than 10-12 paired presentations of a tone and a weak electric shock. These were measured through movement of a stabilimeter (Fig. 4). The purpose was not to determine speed of conditioning but the absolute threshold for a wide range of frequencies.

Not all conditioned responses, however, are formed readily in monkeys. Primates condition rapidly under certain conditions and with great difficulty under others. Conditioning is not an effective measure for distinguishing primates from subprimates, nor lions from lemurs and learned scholars. More effective criteria are the variety and complexity of learning problems which can be mastered. In the first half of the 20th century, literature concerning

Fig. 4. Stabilimeter.

the conditioned response paradigm did not indicate a search for a learning theory applicable to higher organisms but rather a concern of the theorists with rat learning for its own sake. Forty years of intensive research with the conditioned response cast little light on analysis of complex human learning tasks. I believe that there is far greater chance of understanding simple learning through study of the more complex than of understanding complex learning via the simple.

We have been describing simple learning where all an animal learns is to do something better or faster. A more complex kind of learning is discrimination learning, which requires an animal to respond positively to one thing or object and to not respond positively to a different thing or object. Discrimination learning is tested with the WGTA for which several adaptation trials are run prior to testing. The natural response of monkeys is to pull the objects into their cages, but there are techniques to break this response and shape the monkeys into pushing the objects sideways or backwards. One technique is to chain the objects so that they cannot be pulled forward.

Because experimenters are very ingenious in shaping up animals in apparatuses, one might get the impression that any animal can be taught any operant conditioned response in which the correct response is rewarded. Nothing could be further from the truth, as Breland and Breland illustrated with their "piggy bank" pigs. First one coin, then progressively more coins, were placed several feet from a large piggy bank and, after a few trials, a pig would eagerly pick up a dollar, carry it to the bank, be rewarded with food, and then run back for another dollar. A diller, a dollar, a piggy bank scholar. Pigs condition very rapidly, they are naturally ravenous, and they are usually quite tractable. After a period of a week or so, however, the pigs' responses became slower and slower. They would run eagerly for each dollar but on the way back to the bank would drop the dollar, root it, pick it up, toss it in the air, drop it, and root it again. Finally it would take a pig about 10 minutes to transport four coins a distance of 6 feet. He would not receive enough rewarding food to have enough to eat during the day. This problem developed in successive pigs. The pigs failed because of rooting and tossing. They found it more pleasant to bury money than to spend it. Any animal can be trained to do what comes naturally, but no animal can be trained to perform in opposition to his natural tendencies and capabilities.

The simplest kind of discrimination learning is spatial discrimination, in which an animal is trained to go to one place and not to go to another place. This is taught by rewarding the responses made to a particular positional cue when alternate positional cues are present. Primates are trained to select the right or left object of two identical objects covering foodwells on the WGTA tray. As a rule monkeys solve spatial discrimination problems in a single trial. This is called insightful on site learning. Spatial discrimination is simple because there is no ambiguity in the problem. All the animal has to do is to differentiate left from right and right from wrong.

In the problems described thus far, learning may be achieved with little or no interference with the stimulus cues. In simple conditioned response learning only one stimulus, the conditioned stimulus, produces a response that is rewarded. Other stimuli in the situation precede or follow the reinforcement or, if simultaneous, lack reinforcement. In discrimination learning interference is essentially limited to the similarities in physical properties of the stimuli within reasonable range of the rewarded stimulus. In spatial discrimination this interference would result from generalization of the spatial place of the reward. For instance, a host of studies demonstrate that the closer the loci of the stimuli between which the choice is made, the less efficient is the performance.

Object discrimination, next to spatial discrimination in complexity, involves consistently rewarding the monkey for choosing an object with a certain quality, such as a certain form or color or degree of brightness. In object discrimination one must always choose the blue object as opposed to the yellow, the black as opposed to the white, or the tarantula as opposed to the turnip. On any one particular trial, object discrimination presents an ambiguity not present in simple spatial discrimination. There is on this particular trial simultaneously a reward of the correct object quality and the position occupied by the correct object. In order to solve the problem, the animal must learn to consistently respond to the correct color or form or degree of brightness and at the same time learn to ignore the position which is now of no significance. To make sure that the subject is responding to object quality and not to the position of the object, the position of the correct or rewarded stimulus is varied from right to left in random fashion. Learning is dependent upon recognition of a differential in frequency of reward. In simple object discrimination the experiment is so devised that the correct object quality is rewarded 100% of the time and each position only 50% of the time over a series of trials.

Spatial and object learning in real life situations are often so involved with a multiplicity of other factors that the various elements cannot easily be dissociated from each other. When Galahad set out to find the Holy Grail, for instance, it took him longer to find the place and a hell of a lot longer to get back than it did to find the Holy Grail. One would not infer from this, however, that spatial learning was simpler than object learning. More seriously, most seeing animals who can learn spatial, positional discriminations can learn object discriminations at the same time (see Fig. 5). This is fortunate because if this capability did not exist no one would ever find his way in New York City or, what may be even more important, no one would find his way out of New York City.

Monkeys with no previous experience on object discrimination solve these problems speedily, even taking into consideration some incidental learning from the preliminary adaptation trials. Again, this speed of learning does not establish a differentiating learning criterion between primates, human or subhuman and lower families, genera, or species. Rats also learn speedily the nonspatial discrimination between all black and all white alleys of a maze. Experienced monkeys, in addition, solve species of nonspatial problems practically without error after an initial blind trial, an accomplishment not yet equaled by any subprimates. Possibly this ability of immediate or almost immediate solution of a series of nonspatial or object discriminations can differentiate some learning of monkeys and man from that of subprimates.

Fig. 5. Object discrimination.

We devised more complex problems in which there is ambiguous reward of *more than two cues* on any particular trial. Out of these several cues the monkey must learn to respond to only one cue. These problems the author named multiple-sign problems, Lashley called them generalizations of a second or higher order, and Nissen described them as ambivalent cue problems.

There are two illustrations of multiple-sign situations which are similar in the fact that they complement each other. These are the matching and the oddity problems. In the first, the subject chooses from three or more objects the identical objects whereas, in the second, he chooses the odd one out of a group of three or more objects. Monkeys also find these problems comparable and require about the same number of trials for the mastery of each.

Matching was first studied by Kohts (1928) for testing visual capacities of chimpanzees and a variation of her technique was standardized at Wisconsin for monkeys. Winsten (1945) subjected the fundamental procedures to rigid laboratory controls. Winsten became so intrigued by his matching monkeys that he believed the matching test to be as important a scientific

Fig. 6. The oddity problem.

contribution as the discovery of insulin. In view of the fact that we now know that insulin did not solve the intended problem, he may have been right. Winsten achieved remarkable data through intense motivation. He and the monkeys knew each other well through years of living together. When a monkey made a mistake, he slammed the screen door in front of it and shouted "NO!". Each time the monkey barely rescued his hands, looked frightened and stopped responding, then began cooperating again. There is some question as to how thoroughly the monkeys learned the matching problem but no question at all about Dr. Winsten. In the simplest form of matching, the test tray is divided into two compartments, a sample compartment with one foodwell covered by an object and a choice compartment with two or more foodwells, also object covered. The subject is trained to displace the rewarded sample object and then to select the matching object from the choice compartment.

An oddity problem is illustrated in Figure 6. Only three stimuli are presented out of two different pairs of identical stimulus objects used throughout the problem. The simple oddity test was discovered at Wisconsin

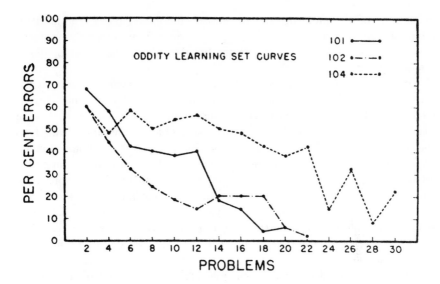

Fig. 7. Individual differences in oddity performance.

some years after it had been discovered at Princeton or the Yerkes Laboratory. It is one of the least important of the Orange Park discoveries, but one of the most important discoveries of Princeton. The odd stimulus object is rewarded and the testing arranged so that one of each pair is odd in half the test trials. Over the series of trials only the singly presented or odd object is rewarded 100% of the time, although on any one trial the position and the object are rewarded as well and must be disregarded as the monkey masters the problem. Figure 7 shows a superior, an average, and a poor performance on an oddity series by three monkeys who have their own individual differences even though they have no report cards to take home.

Some of the learning tests which we adapted for the rhesus have been fulfilling their function by enriching human learning techniques. In 1959 Ellis and Sloan suggested a mental age level of 6 years as the approximate human age when oddity learning could be expected. Recent research by Brown shows that the oddity concept causes considerably more confusion for the mildly retarded child than for the normal child of the *same mental* age (Brown, 1970). Through results of the oddity and object discrimination tests, often administered on the WGTA, an attention deficit has been found to be marked among these retarded children. Martin and Tyrrell (1971), as well as Brown, have suggested this deficit as a primary cause of the contradiction between accomplishments and ability levels in the retarded and normal children. Special cues have been used to focus the child's attention on differentiating dimensions of the problems and to reduce random responses. The need for attention cues increases if the object quality varies from problem to problem, as from color to form, and is greater for the oddity test than for object discrimination.

To make a monkey's life and learning more complicated, the next step in the hierarchy of problems is a combination of matching-nonmatching or oddity-nonoddity, also called third order multiple-sign problems. In the matching-nonmatching, if the sample stimulus object is rewarded with food the monkey is to choose the matching object of the two choice stimulus objects. No food reward under the stimulus object is a sign to the animal to select the object unlike the sample. Four monkeys mastered this problem by first learning matching in 50 trials, then nonmatching, and finally the combination presented in irregular sequence. An average 1900 trials were required for solution, but the stepwise procedure accomplished the feat. This is just about twice the average number of trials for oddity or matching mastered separately.

Next we produced the board sign principle which required the monkey to choose the odd stimuli if they were on a green test board and to choose one of the identical objects if they were on a yellow test board. The heart of this problem is that the stimulus originally nonodd becomes odd, so that the animal must learn to ignore the object as well as to ignore the space. The hierarchy is both logical and factual. We do not have totally comparable human data, but these tests will probably strain the learning ability of human children 7 and 8 years of age or even older. Monkeys are not little men, but they probably do not stand in awe of the little men who are little men.

An adaptation of the Weigl Test (Harlow, 1942, 1943) proved to be an even more provoking 3rd order multiple-sign problem. The test was adapted from the well-known Weigl Test of Abstraction used clinically with human beings. The problem complexity was increased by the use of four stimuli, two of which were alike in color and two in form. When used in matching tests the monkeys not only had to decide to match but also whether to match for color or for form, depending upon the sign given. The sign was the presence or absence of a food reward under the sample. Differential boards gave the clues for oddity. An orange tray was the sign that a color-odd object should be chosen and a cream-colored tray indicated form.

Weigl was a student of Gelb who worked with Goldstein and, some people believed, thought for him as well. Goldstein (1939) was captivated by Weigl's test because he believed there were two kinds of intelligence, abstract and concrete. Goldstein also believed that the Weigl test tested abstract learning, which was strictly a human capability not granted to subhuman primates.

At the Yerkes Laboratory seven chimpanzees were tested on simplified Weigl matching, at Wisconsin four monkeys on the complicated test, but step-wise procedures were used in both investigations. Three of the four monkeys solved all the tests at a significant level in from 2400 to 5600 trials. This may not sound impressive, but the surprising and striking finding was that the performance of the rhesus monkey was superior to that of the chimpanzee. Only one of the seven chimpanzees reached solution at a significant level and she required over 6700 trials.

Subsequently we tried to produce Weigl insight in monkeys where each problem was given for only a very few trials. We produced Weigl type learning but only in the experimenters. In a final attempt to test the complexity of multiple-sign problems within the capacity of the rhesus monkey, the exhaustive tests exhausted both the experimenters and the subjects, but the monkeys did solve the problem in from 2500 to 3800 trials. The final stage of this test involved randomly presented trials for matching and nonmatching plus two types of antagonistic position trials.

Multiple-sign learning injects some insight into the way in which human behavior may be changed through the use of signs and symbols. In oddity, matching, and their complex variations, the meaning of an object *as a sign* changes from problem to problem. In interspersed oddity and matching the meaning of a specific stimulus object changes even within a problem, from trial to trial. The subject is required to make one response to one sign and the opposite response to the same object as the sign changes the situation. Consistency and clarity evolve only as the signs acquire meaning. The context of meaning is reduced to its simplest structure. The most everyday example of signs and symbols in human life is, of course, words. The words "I" and "you" change meaning depending upon the speaker and a child must learn that the speaker is the sign for the meaning. When the child is speaking, "I" is the child. When spoken to, the child is "you." After a soulful struggle "I" becomes "you" and "you" becomes "I" at the proper point. It takes even longer to know that there is a great difference between a girl who is awfully pretty and one who is pretty awful.

There is a vast multitude of variables which may interfere with the orderly, predictable process of learning. Analysis of many types of tasks with many species of animals must be experimentally achieved before the functions or generalization of these variables in specific situations can be even estimated. Without such investigations, theorists overestimate certain classes of variables and neglect others.

Certain interfering factors can be identified along with their roles in the learning tasks which have been presented in this chapter, even though the importance of these factors may change during the solution of the problem. Included are the variables of both stimulus and response preferences, the caprice caused by curiosity, and ambiguity of response reward.

Animals are born with some *stimulus preferences* and constantly collect more in the learning process. Evidence for certain innate color preferences, for instance, is shown in the universality of preferences of similar hues and intensities among human beings of different races and cultures. In many countries man dislikes the color black as a choice for colors in flags, and, even if monkeys have not yet chosen their respective flags, they don't like black either. If a blue and a black object are presented to a rhesus monkey in an object discrimination problem, the monkey will first choose the blue. If the blue is rewarded on the first choice, the problem presents no difficulty, but, if the unpreferred black is rewarded, learning will be definitely retarded. In our discrimination problems, stimulus-preference errors were frequent in the early problems, but also were the first kind of error to be eliminated.

Not surprisingly, there are also preferred *response preferences* for behavior which is more pleasant and, therefore, is reinforced more readily. Monkeys would rather pull a plunger out than push a plunger in. If pulling a plunger out is the correct response for opening a problem box, learning will be facilitated. If one has ever tried to push a pile of leaves or grass with a rake it will be easy to understand a response preference shared by monkeys, gibbons, chimpanzees, and human children and adults. Pulling in a rake is the preferred response, with a sidewise motion next, and pushing objects away the least preferred motion. In testing tool using, both the animals and the children will learn with the greatest of ease if food is placed near the angle of the handle and the blade of the rake where it can be pulled toward the subject. An example of a learned response preference is the tendency to choose the stimulus object just recently reinforced, prior to a discrimination reversal.

Working at times counter to the choice of a preferred stimulus is the constant curiosity and urge to explore, an extremely persistent peculiarity of both monkey and child. There may, suddenly, after a successive series of correct and rewarded choices, be what looks like a spontaneous shift of response on the part of the subject. The monkey or child just cannot resist the urge to find out what lies beneath the second stimulus, which has suddenly become seductive just because it conceals an unexplored unknown. This factor, which we named the response shift, is probably more active among primates than among lower animals, such as the rat and the robot, which seem to be satisfied more readily with monotony.

In discussing the more complex learning problems, the question of *ambiguity of reward* has been discussed as a factor increasing difficulty of solution. In the first correct choice in an object discrimination, both the position and the object may be considered by the subject to be rewarded. Not until the correct object is rewarded in a different position will this ambiguity begin to disappear. Similarly, a child whose behavior is negatively reinforced in an ambiguous manner will find it more difficult to learn. If father says "Stop that." when Tommy is both laughing and pulling the kitten's tail, it may take longer to learn whether he is to stop laughing or to stop pulling the tail.

In the development of primate learning tests, a fundamental principle underlying relationships between stimuli and responses was found to be operative during the learning process. This principle is called *stimulus generalization*. If a subject has become familiar with one particular stimulus object and, in a new test, a second stimulus object which differs in only one single dimension is introduced, for instance a smaller blue circle, the monkey will still choose the original stimulus. The closer in size the new circle is to the original, however, the less frequently the animal will choose the original, and the opposite is also true. The further apart in size is the new circle, the more frequently the animal will choose the original stimulus. There has occurred generalization of response to stimuli similar to the original stimulus object. Bromer (1940) found that generalization occurred in oddity as well as in discrimination problems, but, oddly enough, there was greater generalization of the oddity response to color than to form. Numerical generalization, as from a three-foodwell board to four and five-foodwell trays, was demonstrated in both the simple oddity and in the Weigl-oddity problems.

Having created an adequate number of distinct learning tests differing both qualitatively and quantitatively, we then proceeded to use these in the analysis and evaluation of basic psychological problems. One of the first was the maturation of learning, since many annectant problems relate to animal age and learning capability. We studied the development of learning in the rhesus monkey and found the maturational age at which various problems could be mastered. Monkey intelligence evolves just like human intelligence and we created a monkey mental age scale, less efficient than the Binet, but more efficient than any predecessor of the Binet-Simon test.

We also studied the effect on learning of depriving the animal of any voluntary activity. We used two different techniques. The first was complete striate muscular paralysis produced by curare. The other was an ingenious method of paralyzing all movement in the conscious animal by inactivating the motor and somesthetic cortex by application of ethyl chloride. These studies clearly demonstrated that learning is not contingent upon bodily movements. This fact has subsequently been shown to be true for man.

The fact that such varied and intricate learning functions can be achieved so early and efficiently attests to the tremendous learning ability in all infant primates far beyond the complexity of imprinting. Thus complex yearning is gradually overwhelmed by complex learning. Had we not previously engaged in countless years of monkey learning studies, we would not have discovered the major relation between innate and learned variables. In the primates—monkeys, apes, and men—learning is a variable of such power and promise that as soon as any innate behavior appears it is covered and concealed by immediate learning and even by antecedent learning. If one is going to study behavior in primates and in all animal forms, one should know the capabilities of learning and the time of the sequence of maturation of behaviors at all levels of complexity.*

*For further detailed data the reader is referred to Harlow, H. F., Primate learning. In C. P. Stone (Ed.), *Comparative psychology*, from which large excerpts have been reprinted by permission. Copyright 1951 by Prentice-Hall.

CHAPTER 2

THE DEVELOPMENT OF LEARNING

During the 1950s we conducted an integrated series of researches tracing and analyzing the learning capabilities of rhesus monkeys from birth to intellectual maturity. Control over the monkey's environment has been achieved by separating the infants from their mothers at birth and raising them independently, using techniques and methods adapted from those described by van Wagenen (1950).

The magnificent, marvelous macaque monkey has a predetermined maturation rate, proportional to that of the human being. Monkeys learn problems that are simple and problems that are complex. Furthermore, their efficiency in solving simple problems increased with increasing age and the complexity of learning problems that monkeys can master is a function of chronological concordance.

There are many characteristics that commend the rhesus monkey as a subject for investigation of the development of learning. At birth, or a few days later, this animal attains adequate control over its head, trunk, arm, and leg movements, permitting objective recording of precise responses on tests of learning. The rhesus monkey has broad learning abilities, and even the

27

neonatal monkey rapidly learns problems appropriate to its maturation status. As it grows older, this monkey can master a relatively unlimited range of problems suitable for measuring intellectual maturation. Although the rhesus monkey matures more rapidly than the human being, the time allotted for assessing its developing learning capabilities is measured in terms of years—not days, weeks, or months, as is true with most subprimate forms. During this time a high degree of control can be maintained over all experimental variables, particularly those relating to the animal's learning experiences. Thus, we can assess for all learning problems the relative importance of nativistic and experiential variables, determine the age at which problems of any level of difficulty can first be solved, and measure the effects of introducing such learning problems to animals before or after this critical period appears. Furthermore, the monkey may be used with impunity as a subject for discovering the effects of cerebral damage or insult, whether produced by mechanical intervention or by biochemical lesions.

The only other creature whose intellectual maturation has been studied with any degree of adequacy is the human child, and the data from this species attest to the fact that learning capability increases with age, particularly in the range and difficulty of learned tasks which can be mastered. Beyond this fact, the human child has provided us with astonishingly little basic information on the nature or development of learning. Obviously, there are good and sufficient reasons for any and all such deficiencies. There are limits beyond which it is impossible or unjustifiable to use the child as an experimental subject. The education of groups of children cannot be hampered or delayed for purposes of experimental control over either environment or antecedent learning history. Unusual motivational conditions involving either deprivation or overstimulation are undesirable. Neurophysiological or biochemical studies involving or threatening physical injury are unthinkable.

Even aside from these cultural limitations, the human child has certain characteristics that render him a relatively limited subject for the experimental analysis of the maturation of learning capability. At birth, his neuromuscular systems are so undeveloped that he is incapable of effecting the precise head, arm, hand, trunk, leg, and foot movements essential for objective measurement. By the time these motor functions have adequately matured, many psychological development processes, including those involving learning, have appeared and been elaborated, but their history and nature have been obscured or lost in a maze of confounded variables.

By the time the normal child has matured physically, he is engaging each day in such a fantastic wealth of multiple learning activities that precise, independent control over any single learning process presents a task beyond objective realism. The multiple, interactive transfer processes going on overwhelm description, and their independent experimental evaluation cannot be achieved. Even if it were proper to cage human children willfully, which it assuredly is not, this very act would in all probability render the children abnormal and untestable and again leave us with an insuperable problem.

It might appear that all these difficulties could be overcome best by studying the development of learning abilities in infraprimate organisms rather than monkeys. Unfortunately, the few researches which have been completed indicate that this is not true. Animals below the primate order are intellectually limited compared with monkeys, so that they learn the same problems more slowly and are incapable of solving many problems that are relatively easily mastered by monkeys. Horses and rats, and even cats and dogs, can solve only a limited repertoire of learning tasks, and they learn so slowly on all but the simplest of these that they pass from infancy to maturity before their intellectual measurement can be completed. Even so, we possess scattered information within this area. We know that cats perform more adequately on the Hamilton perseverance test than do kittens, and that the same relationship holds for dogs compared with puppies (Hamilton, 1911, 1916). It has been demonstrated that mature and aged rats are no more proficient on a multiple-T maze than young rats (Stone, 1929), and that conditioned responses cannot be established in dogs before 18–21 days of age (Fuller, Easler, & Banks, 1950); but such data will never give us insight into the fundamental laws of learning or maturation of learning.

The most surprising finding relating to neonatal learning was the very early age at which simple learning tasks could be mastered. Indeed, learning of both the simple conditioned response and the straight runway appeared as early as the animal was capable of expressing it through the maturation of adequate skeletal motor responses. Thus, we can in no way exclude the possibility that the monkey at normal term, or even before normal term, is capable of rapidly forming simple associations.

Equally suprising is the fact that performance may reach or approach maximal facility within a brief period of time. The 5-day-old monkey forms conditioned reflexes between tone and shock as rapidly as the year-old or the adult monkey. The baby macaque solves the simple straight-alley problem as soon as it can walk, and there is neither reason nor leeway for the adult to do appreciably better. Although we do not know the minimal age for solution of the Y-maze, it is obviously under 15 days. Such data as we have on this problem indicate that the span between age of initial solution and the age of maximally efficient solution is brief. One object discrimination, the differentiation between the total-black and total-white field, shows characteristics similar to the learning already described. The developmental period for solution lies between 6 and 10 days of age, and a near maximal learning capability evolves rapidly. However, it would be a serious mistake to assume

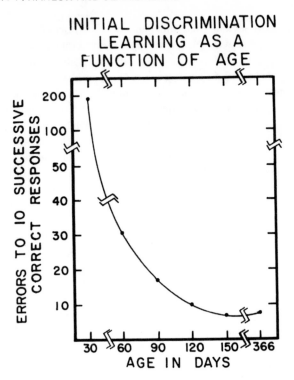

INITIAL DISCRIMINATION LEARNING AS A FUNCTION OF AGE

Fig. 8. Initial discrimination learning as a function of age.

that any sharply defined critical periods characterize the development of more complex forms of learning or problem solving.

Object Discrimination Learning. Although the 11-day-old monkey can solve a total-black vs. total-white discrimination problem in less than 13 trials, the 20- to 30-day-old monkey may require from 150 to 200 trials to solve a triangle-circle discrimination problem when the stimuli are relatively small and placed some distance apart. It is a fact that, even though the capability of solving this more conventional type of object-discrimination problem exists at 20 days, object-discrimination learning capability has by no means attained full maturity at this time.

The development of complete object-discrimination capacity was measured by testing five different age groups of naive rhesus monkeys on a single discrimination problem. Discrimination training was begun when the animals were 60, 90, 120, 150, or 366 days of age, and, in all cases, training was preceded by at least 15 days of adaptation to the apparatus and to the eating of solid food.

Figure 8 presents the number of trials taken by the five different groups of monkeys, and performance by a 30-day-old group on a triangle-circle discrimination in plotted on the far left. Whether or not one includes this group, it is apparent that the ability of infant monkeys to solve the object-discrimination problem increases with age as a negatively accelerated function and approaches or attains an asymptote at 120 to 150 days.

Detailed analyses have given us considerable insight into the processes involved in the maturation of this learning function. Regardless of age, the monkeys' initial responsiveness to the problem is not random or haphazard. Instead, almost all the subjects approached the problem in some systematic manner. About 20% of the monkeys chose the correct object from the beginning and stayed with their choice, making no errors! Another 20% showed a strong preference for the incorrect stimulus and made many errors. Initial preference for the left side and for the right side was about equally frequent, and consistent alternation-patterns also appeared. The older, and presumably brighter, monkeys rapidly learned to abandon any incorrect response tendency. The younger, and presumably less intelligent, monkeys persisted longer with the inadequate response tendencies, and very frequently shifted from one incorrect response tendency to another before finally solving the problem. Systematic responsiveness of this type was first described by Krechevsky (1932) for rats and was given the name of "hypotheses." Although this term has unfortunate connotations, it was the rule and not the exception that our monkey subjects went from one "hypothesis" to another until solution, with either no random trials or occasionally a few random trials, intervening. The total number of incorrect, systematic, response tendencies before problem solution was negatively correlated with age.

These data on the maturation of discrimination learning capability clearly demonstrate that there is no single day of age nor narrow age-band at which object-discrimination learning abruptly matures. If the "critical period" hypothesis is to be entertained, one must think of two different critical periods, a period at approximately 20 days of age, when such problems can be solved if a relatively unlimited amount of training is provided, and a period at approximately 150 days of age, when a full adult level of ability has developed.

Delayed Response. The delayed-response problem has challenged and intrigued psychologists ever since it was initially presented by Hunter (1913). In this problem the animal is first shown a food reward, which is then concealed within, or under, one of two identical containers during the delay period. The problem was originally believed to measure some high-level ideational ability or "representative factor"—a capacity that presumably

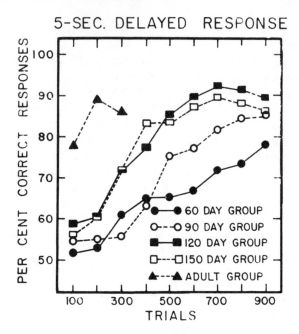

Fig. 9. Delayed-response learning as a function of age.

transcended simple trial-and-error learning. Additional interest in the problem arose from the discovery by Jacobsen (1935) that the ability to solve delayed-response problems was abolished or drastically impaired by bilateral frontal lobectomy in monkeys.

Scores of researches have been conducted on delayed-response, problems. Almost all known laboratory species have been tested and all conceivable parameters investigated. Insofar as the delayed response is difficult, it appears to be less a function of period of delay or duration of memory than an intrinsic difficulty in responding attentively to an implicit or demonstrated reward. However, in spite of the importance of the problem and the vast literature which has accumulated, there has been no previous major attempt to trace its ontogenetic development in subhuman animals.

Ten subjects in each of four groups, a 60-, 90-, 120-, and 150-day group, were tested on so-called zero-second and 5-second delayed responses (the actual delay period is approximately 2 seconds longer) at the same time they began their discrimination learning.

The results for the four infant groups on the 5-second delayed responses and the performance of a group of adults with extensive previous test experience on many different problems are presented in Figure 9. The four infant groups show increasing ability to solve delayed responses both as a function of experience and as a function of age. The performance of all infant groups is inferior to the adult group, but differences in past learning experiences preclude any direct comparison.

The performance of the four infant groups of monkeys on the 5-second delayed responses for trials 1–100, 201–300, 401–500, 601–700, and 801–900 is plotted in Figure 10. Because performance during trials 1–100 is poor regardless of group, it is apparent that a certain minimum experience is required to master the delayed-response task. At the same time, the increasingly steep slopes of the learning curves make it apparent that efficiency of delayed-response learning and performance is in large part a function of age. The group data suggest that, after extensive training as provided in this study, 70% correct responses may be attained by 150 days of age, 80% by 200 days, and 90% by 250 days.

Very marked individual differences were disclosed during delayed-response testing, a finding which typifies this task regardless of species or age. Some monkeys, as well as some other animals, fail to adapt to the requirements of the test. Inspection of individual records reveals that the capability of solving

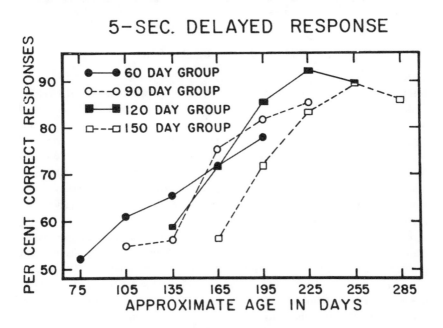

Fig. 10. Delayed-response performance of different maturational groups with age held constant.

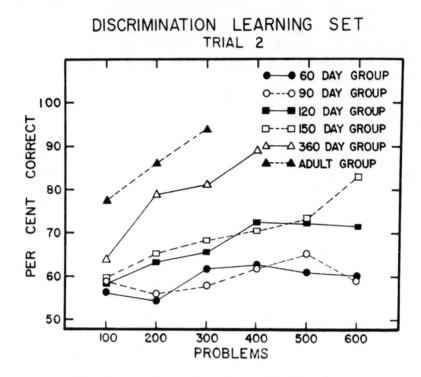

Fig. 11. Learning set formation as a function of age.

this problem first appears at about 125 to 135 days of age and that essentially faultless performance may appear by 200 to 250 days in perhaps half the monkey subjects. Thus, some monkeys at this age may possess an adult capability, and these data are in keeping with the results obtained in the Series II tests. Recently, we have completed a study on a 30-month-old group of five monkeys on zero- and 5-second delayed responses, and their learning rates and terminal performance are at adult levels. Thus, it appears that we have definitive data on the maturation and acquisition of delayed-response performance by rhesus monkeys.

It is obvious that the capability of solving the delayed-response problem matures at a later date than the capacity of solving the object-discrimination problem. This is true regardless of the criterion taken, whether it is the age at which the task can be solved after a relatively unlimited number of trials or the age at which a full adult level of mastery is attained. At the same time, it should be emphasized that this capacity does develop when the monkey is still an infant, long before many complex problems can be efficiently attacked and mastered. Thus, there is no reason to believe that the delayed response is a special measure of intelligence or of any particular or unusual intellectual function.

The same five infant groups previously tested on a single object-discrimination problem served as subjects for learning-set training. The Trial 2 performance of the five groups of infant monkeys is plotted in Figure 11, and data from mature monkeys tested in previous experiments are also given. The two younger groups fail to respond consistently above a 60% level, even though they were approximately 10 and 11 months of age at the conclusion of training. The two older groups show progressive, even though extremely slow, improvement in their Trial 2 performances, with groups 120 and 150 finally attaining a 70 and 80% level of correct responding. These data are in general accord with those obtained from an earlier, preliminary experiment and indicate that the year-old monkey is capable of forming discrimination learning sets even though it has by no means attained an adult level of proficiency.

In Figure 12, the Trial 2 learning-set performance for the various groups is plotted in terms of age of completion of consecutive 100-problem blocks, and these data suggest that the capacity of the two younger groups to form discrimination learning sets may have been impaired by their early, intensive

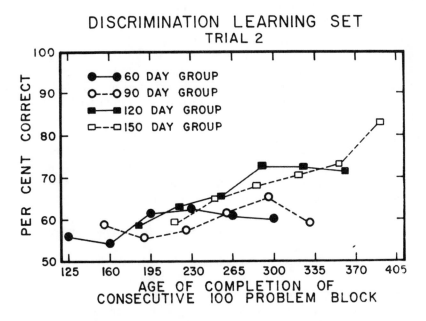

Fig. 12. Learning set plotted for age of completion of consecutive 100 problem blocks.

learning-set training, initiated before they posessed any effective learning-set capability. Certainly, their performance from 260 days onward is inferior to that of the earlier groups with less experience but matched for age. The problem which these data illustrate has received little attention among experimental psychologists. There is a tendency to think of learning or training as intrinsically good and necessarily valuable to the organism. It is entirely possible, however, that training can either be helpful or harmful, depending upon the nature of the training and the organism's stage of development.

Because of the fundamental similarities existing between the learning of an individual problem and the learning of a series of problems of the same kind, it is a striking discovery that a great maturational gulf exists between efficient individual-problem learning and efficient learning-set formation. Information bearing on this problem has been obtained through detailed analyses by the author. The author's error-factor analysis technique reveals that, with decreasing age, there is an increasing tendency to make stimulus-perseveration errors, i.e., if the initially chosen object is incorrect, the monkey has great difficulty in shifting to the correct object. Furthermore, with decreasing age there is an increasing tendency to make differential-cue errors, i.e., difficulty in inhibiting, on any particular trial, the ambiguous reinforcement of the position of the stimulus which is concurrent with the reinforcement of the object per se.

In all probability, individual-problem learning involves elimination of the same error factors or the utilization of the same hypotheses or strategies as does learning-set formation. However, as we have already seen, the young rhesus monkey's ability to suppress these error factors in individual-problem learning does not guarantee in any way whatsoever a capacity to transfer this information to the interproblem learning-set task. The learning of the infant is specific and fails to generalize from problem to problem, or, in Goldstein's terms, the infant possesses only the capacity for concrete thinking. Failure by the infant monkey to master learning-set problems is not surprising inasmuch as infraprimate animals, such as the rat and the cat, possess the most circumscribed capabilities for these interproblem learnings, and it is doubtful if most birds or bats possess any such ability at all. Indeed, discrimination learning-set formation taxes the prowess of the human imbecile and apparently exceeds the capacity of the human idiot.

At the present time we have completed a series of experiments which clearly demonstrate that the capability of solving problems of increasing complexity develops in rhesus monkeys in a progressive and orderly manner throughout the first year of life. Furthermore, when we compare the performances of the year-old monkey and the adult monkey, it becomes obvious that maturation is far from complete at the end of the first year.

Fig. 13. Hamilton perseverance test.

Although our data on early development are more complete than our data on terminal learning capacities, we have already obtained a considerable body of information on middle and late learning growth.

Hamilton Perseverance Test. Just as we were surprised by the delayed appearance of the capability of mastering the patterned-strings tests, so were we surprised by the delay before performance on the Hamilton perseverance test (Fig. 13) attains maximal efficiency.

Three groups of monkeys were initially tested at 12, 30, and 50 months of age, respectively. The groups comprised six, five, and seven monkeys, and all were tested 20 trials a day for 30 days. On the perseverance problem the animal is faced with a series of four boxes having spring-loaded lids which close as soon as they are released. Only one box contains food, and the rewarded box is changed in a random manner from trial to trial with the provision that the same box is never rewarded twice in succession. In the present experiment the subjects were allowed only four responses per trial, whether or not the reward was obtained, and an error was defined as any

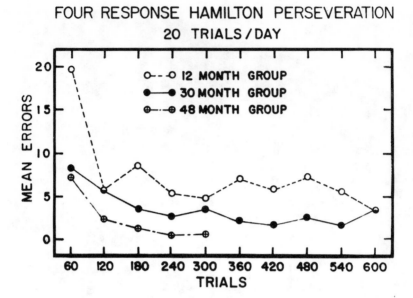

Fig. 14. Mean number of errors per 60-trial block on Hamilton perseverance test.

additional response to an unrewarded box after the initial lifting of the lid
during a trial. Infraprimate animals make many errors of this kind, but as
presented in Figure 14, the mature monkey makes few such errors and learns
rather rapidly to eliminate these. We were surprised by the inefficient
performance of the year-old monkey and unprepared to discover that
maximally efficient performance was not attained by the 30-month-old
monkeys.

The mature monkey finds a simple plan for attacking the perseverance
problem. Typically, it chooses the extreme left or right box and works
systematically toward the other end. If it adopts some more complex strategy,
as responding by some such order as box 4-2-3-1, it will repeat this same
order on successive trials.

Since the animal's procedural approach to the perseveration problem
appeared to be an important variable, measures were taken of changes in the
animal's order of responding from trial to trial, and these were defined as
response-sequence changes. The data of Figure 15 show that the
50-month-old monkeys adopt the invariant type of behavior described above
but that this is not true for either the 12- or 30-month-old groups. If a
subject adopts an invariant response pattern, the problem is by definition
simple; failure to adopt such a pattern can greatly complicate the task. In

FOUR RESPONSE HAMILTON PERSEVERATION
20 TRIALS / DAY

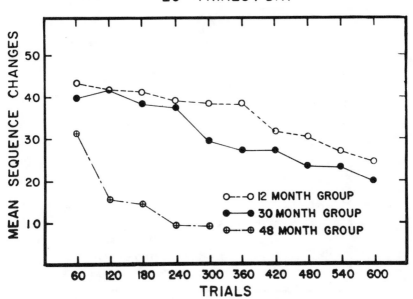

Fig. 15. Mean number of sequence changes per 60-trial block on Hamilton perseverance test.

view of this fact it is not surprising that the 30-month-old subjects made so many errors; rather, it is surprising that they made so few—their error scores represent a triumph of memory over inadequate planning.

Relatively little research on the Hamilton perseverance method has been conducted by psychologists in spite of the fact that the original studies resulted in an effective ordering of Hamilton's wide range of subjects in terms of their position within the phyletic series. Furthermore, the limited ontogenetic material gave proper ordering of animal performance: Kittens, puppies, and children were inferior to cats, dogs, and human adults. Above and beyond these facts, the perseverance data give support to the proposition that the rhesus monkey does not attain full intellectual status until the fourth or fifth year of life.

We have now completed a series of experiments using the two-position oddity problem, in which the correct stimulus is either in the right or left position, never in the center. In Figure 16 are presented data from a group of 10 monkeys tested on 256 problems at 20 months of age and again at 36

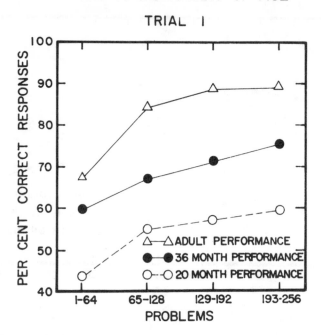

Fig. 16. Performance on oddity problems as a function of age.

months of age. These data indicate increasingly efficient performance as a function of age.

Additional oddity learning data was obtained by the same training techniques. Again a group of 10 monkeys was trained on oddity, first at 12 months of age and subsequently at 36 months of age. At 30 months of age, however, this group was divided into two groups of five monkeys each, one group being trained on a series of 480 six-trial discrimination problems and the other 2400 delayed response trials. The differences between the two groups on the oddity problems are statistically significant, with every indication of negative transfer from the learning-set to the oddity training. This is consistent with the fact that a single stimulus is uniformly correct on every discrimination problem but reverses frequently during the trials of each oddity problem.

For the neonatal and infant rhesus monkey, each learning task is specific unto itself, and the animal's intellectual repertoire is composed of multiple, separate, and isolated learned experiences. With increasing age, problem isolation changes to problem generalization, and this fundamental reorganization of the monkey's intellectual world apparently begins in its second year of life. From here on, we can no longer specify the monkey's learning ability for any problem merely in terms of maturational age and individual differences. The variable of kind and amount of prior experience must now be given proper value. This is the characteristic of monkey learning, and, in fact, learning by all higher primates as they come of intellectual age.

The monkey is capable of solving simple learning problems during the first few days of life and its capability of solving ever increasingly complex problems matures progressively, probably for four to five years. Early in life, new learning abilities appear rather suddenly within the space of a few days, but, from late infancy onward, the appearance of new learning powers is characterized by developmental stages during which particular performances progressively improve. There is a time at which increasingly difficult problems can first be solved, and a considerably delayed period before they can be solved with full adult efficiency.

The monkey possesses learning capacities far in excess of those of any other infrahuman primate, abilities probably comparable to those of low-level human imbeciles. The monkey's learning capabilities can give us little or no information concerning human language, and only incomplete information relating to thinking. These are the generalizable limits of learning research on rhesus monkeys, but they still leave us with an animal having vast research potentialities. There is a wealth of learning problems which the monkey can master, and at present the field is incompletely explored. The maturation of any learning function can be traced and the nature and mechanisms underlying interproblem and intertask transfer can be assessed. There exist great research potentialities in analyzing the fundamental similarities and differences among simple and complex learnings within a single species. The monkey is the subject ideally suited for studies involving neurological, biochemical, and pharmacological correlates of behavior. To date, such studies have been limited to adult monkeys, or monkeys of unspecified age, but such limited researches are no longer a necessity. We now know that rhesus monkeys can be raised under completely controlled conditions throughout a large part, and probably all, of their life span, and we may expect that the research of the future will correlate the neurophysiological variables, not with the behavior of the static monkey, but with the behavior of the monkey in terms of ontogenetic development.*

*The authors wish to express appreciation for the use of large excerpts from Harlow, H. F., The development of learning in the rhesus monkey. *American Scientist*, 1959, 47, 459–479. *Reprinted by permission.*

CHAPTER 3

THE EVOLUTION OF LEARNING

There must be few problems that are more difficult of solution than the problem of the evolution of learning. One difficulty lies in the fact that learning leaves no fossils or remains until we have the scattered indirect archeological evidences associated with the rise of man. Fortunately, however, there still exist representatives of most of the major adaptive types—phyla and classes—providing us with a wide range of animals that may be subjected to psychological test.

Another difficulty lies in existing limitations to a precise classification of the various forms of learning and learning problems into levels of difficulty. Although considerable progress has recently been made along these lines, the puzzle is far from solved; and no one has even attempted to scale the various learning problems or classes of problems in steps of equal difficulty. Furthermore, it is hazardous to compare learning ability among animals independently of their sensory and motor capacities and limitations. Diversity in receptor-effector mechanisms frequently renders exact comparisons of learning between species and genera questionable and poses major problems when we attempt comparisons among orders, classes, and phyla.

The existing psychological-biological data make it quite clear that evolution has resulted in the development of animals of progressively greater potentialities for learning and for solving problems of increasing complexity. No one would question that the problem-solving abilities of man exceed those of the amoeba, even though differences in receptor-effector mechanisms preclude the possibility of direct test. Furthermore, it appears that there have been prolonged geologic periods during which progressively improved learning evolved, and this can be indirectly demonstrated with existing representatives of selected orders, classes, and phyla.

Among the sea-dwelling animals, the scanty data which are existent suggest that the primitive Metazoa, such as some coelenterates (Fleure & Walton, 1907) and echinoderms (Maier & Schneirla, 1935), may have attained an ability to profit from experience that exceeded any capabilities developed in the Protozoa, such as the amoeba (Mast & Pusch, 1924) and paramecium (French, 1940; Gelber, 1952). Such evidence as we have on the flatworm (Hovey, 1929) suggests that it stands out as an intellectual giant compared with the most recondite of the Coelenterata (Fleure & Walton, 1907), even though we run the risk that we are looking at intellectual evolution from a worm's eye view. Representatives of the class Oligochaeta of the phylum Annelida (Robinson, 1953; Yerkes, 1912) apparently greatly extended their mental horizons, but the scanty evidence available suggests that they did not attain the level of rationality which we must accord one of the Cephalopoda, the octopus (Boycott & Young, 1950; Schiller, 1949). The learned performances of some of the teleost fishes (Reeves, 1919) probably surpass those of any other nonmammalian marine organism, and it is regrettable that no member of the class Chondrichthyes or the superorder Chondrostei has been subjected to intensive psychological test. Man's interest in the shark has not yet extended to its behavior under higher cerebral control, and the psychology of the sturgeon remains a virgin area for investigation.

The advent of land animals did not result in any sudden superiority of learning capabilities of land over sea vertebrates insofar as can be demonstrated by testing existing representatives of either the Amphibia or the Reptilia. It is doubtful if any amphibian or reptile has demonstrated more complex learning than that exhibited by many teleost fish. This, of course, does not mean that representatives of these two classes did not show adaptive evolutionary changes from those of their fish ancestors, probably the Crossopterygii. It merely implies that there is no conclusive evidence of differential rate in the evolutionary development of learning between the land and the sea vertebrates for the 200 million years extending from the early Devonian to the end of the Cretaceous periods. Some skepticism, however, may be attached to this position in view of the greater development of the cerebral structures in some of the reptiles and the evidence for the beginning of the formation of the cerebral cortex. It is possible that our behavior tests on the turtle have never done justice to it, and that a latent imagination has been obscured by an introverted personality.

It would be an error, of course, to assume that the evolution of intelligence was dormant or delayed from the Cambrian to the Devonian period. During this interval of 200 million years evolution may have progressed from the Echinodermata, which must surely be able to learn even though they have resisted the attempts of biologists and psychologists to demonstrate clear-cut learning (Maier & Schneirla, 1935), to some shark-like form which probably learned some simple problems with ease. Within another 100 million years the goldfish and the turtle evolved, and with them came capability of learning a reasonable range of problems. During the next 100 million years the mammals and the birds may or may not have been making progress in the evolution of learning ability, but progress is by no means unlikely in view of the clear-cut advances that were to be made in the next, and last, 75 million years. It is important to bear in mind that we have no way to scale, in anything approaching equal-step intervals, the difficulty of various classes of problems, and until this can be achieved on some basis other than intuition and anthropocentrism we cannot judge whether or not the evolution of learning capability has increased or decreased in rate in any 100-million-year block of time, including the last. Whether or not the evolution of learning has increased more rapidly in the last 100 million years than in the 100-million-year blocks preceding, the evolution has been of a nature that greatly simplifies the psychologists' problems of testing and evaluation.

It is interesting to look at the evolution of learning from the anatomical point of view. We make the assumption that learning is primarily a function of the nervous system, or at least that complexity of learning is intimately related to the developing complexity of the nervous system. If we were to examine learning, using the same kind of evidence that we use for assessing locomotion in the evolution of the horse—the anatomical record—we would be struck by a number of facts. Between the Protozoa and the Coelenterata there must be a vast evolutionary gulf, for the members of the one phylum possess no nerve cells whereas the members of the other do. Between the Coelenterata and Platyhelminthes there must be another separation, but one of lesser magnitude. In both kinds of organisms the mechanisms associated with coordination and adjustment is neural, but in the flatworm we find a new kind of organized structure, the cephalic ganglion, and this particular structure and its elaborations are going to characterize all higher nervous systems from here through man. Between the flatworm and the dogfish there is also a gulf, but a lesser one that either of the two previous separations; in this instance we have more neurons and some relatively small changes in their physical elaborations in the forebrain, suggesting an increasing differentiation of their single, basic function. From dogfish to man the separation is very slight: The number of neurons has increased, and the process of structural differentiation has continued. From monkey to man there is essentially no difference other than a very slight tendency to continue the evolutionary trends previously noted.

The very striking fact is that the anatomical record of evolution of the nervous system, including the brain and cerebral cortex, is a continuous and highly orderly process, and there is no evidence that the developmental rate ever suddenly increased, certainly not in the last million years, nor in the last 20 million years, the last 200 million years, or from the beginning of life. The behavioral point of view appears to differ from the anatomical, at least from the human standpoint. Because I have vast respect for anatomy, it is my prejudice that the anatomical point of view is correct, and that, as we become more and more sophisticated concerning the relative difficulty of kinds of learning problems, the learning data relating to evolution will come into accord with those of the anatomy of the development of the nervous system.

Remaining, for the time being, in the land of speculation, I would like to hazard certain guesses as to how learning evolved. It is my understanding that evolution operates through the selection of different genotypes which differ in their ability to produce adaptive responses to particular environmental changes, and that the dynamic forces which produce change in gene frequency are mutation, selection, migration, and genetic drift.

One may seriously ask, "What is it about learning or improved learning that confers upon organisms some slight advantage which has made possible the selective changes in natural populations that have led to the evolution of the remarkable learning capabilities that characterize the order of primates?" Because we look at the world from a Homo sapiocentric point of view, we may attach such importance to learning that we accept as an axiom that learning, all learning, is good and should therefore have survival value. Furthermore, we can all cite many examples in which it would appear self-evident that learning has survival value for the organism. The ability of the sea anemone gradually to differentiate between food and nonfood could confer upon it a slight survival advantage over some other oganism of less brilliant intellectual endowments. The capacity to form conditioned escape responses from noxious and dangerous stimuli would appear to have obvious utility even if, by virtue of the slowness of learning, 99% of the organisms perished before the conditioned response was established. The acquired gift of manufacturing weapons is an obvious selective biological gain and appears to have greatly aided a particular species in survival—at least so far.

I have long been puzzled by the fact that the study of animals under laboratory conditions reveals many learning capabilities whose existence is hard to understand in terms of survival value. Previous discussions of this problem has engendered considerably more heat than light. As an example, the earthworm can learn a spatial maze, i.e., it can learn eventually to turn right for the reward of a bed of succulent mud and not turn left because of the threat of shock or sandpaper. Under the most idealized laboratory conditions, the earthworm solves this task in a faltering and ephemeral way in a few hundred trials (Robinson, 1953; Yerkes, 1912). For more primitive organisms, this is a learning landmark; but even so it it hard to see how this feat of learning legerdemain aided the earthworm, or any other animal so endowed, to survive at the expense of less gifted associates.

It might be argued that the earthworm's learning in nature is more efficient than in the laboratory, or that its limited learning is peculiarly adapted to its natural environment and for this reason provides some evolutionary gain. But this is pure speculation, and there are not even the hopelessly inadequate data, which naturalists so commonly and gladly provide, to give factual support to any such position. The best, not the worst, annelidan learning has been observed by scientists.

I was puzzled for many years as to how the rhesus monkey or chimpanzee developed the capability of solving the complex multiple-sign or conditional problems, including oddity, matching, Weigl-type matching, or the categorization of kinds of stimuli into classes of forms, colors, or—within limits—number (Hicks, 1956; Weinstein, 1955). The observational accounts of these animals make it quite clear that problems of this level of complexity are never solved, indeed, they are never met, in the natural environment. It is superficially difficult to see how a trait which was never used gave to an organism some slight selective advantage over another organism which did not use the trait because it did not have it. Yet such capabilities must have existed in the prehuman primate for some millions of years before the organism developed to the point at which it could put these traits to effective use and convey a clear selective advantage to man.

It can again be argued that those learning capacities essential for complex color categorization by monkeys and apes are used by the same animals in the wild in some manner to provide selective advantage. If this is true, it is something which has eluded, or not been reported or recognized by, the well-trained psychologists and biologists who have gazed patiently at the unending ingenuousness of the social lift of monkeys and apes and their adaptations to nature (Bingham, 1932; Carpenter, 1934, 1942a, 1942b; Nissen, 1931).

Since we must accept as fact that evolution is orderly and results from the selection of gradual changes in the gene population, and since an explanation in terms of autogenesis in untenable, it is interesting to speculate how receptive and neural mechanisms underlying learning might have developed and provided selective gain during the process of evolution. It is obvious that the explanation we desire is in terms of orthoselection, and any reasonable explanation of the remarkably orderly and prolonged evolutionary development of learning in whole or in part in terms of allometry should be given full consideration.

That there is an intimate relation between the development of the receptors and the development of the central nervous system may be taken as fact. Indeed, if Parker is correct, the nerve fiber and neuron may have evolved from the primitive receptor cell. The developmental status of the various receptor systems in animals clearly puts a limit upon all animals' learning capabilities, and it is a safe generalization that, from the phyletic point of view, learning potentialities always lag behind receptor potentialities. Thus, there may be fish, and there are likely reptiles, and there are certainly many birds and mammals that have the receptor potentialities to convey all environmental information essential for human-type thought. Yet this particular learning capacity, doubtless like countless other learning capacities, lagged far behind the receptor system's resolving power.

From the point of view of natural selection, it is difficult to think of any kind of receptor development which would not result in some selective gain so long as the animal's environment provides an adequate stimulus. Any receptor development, in and of its very nature, demands the development of increasingly complex neuronal systems within the central nervous system and even within the receptor itself. As long as increasingly complex receptor systems provide the organism with slight survival advantages, one can be assured that increasingly complex nervous systems will develop; and as long as increasingly complex nervous systems develop, the organism will be endowed with greater potentialities which lead inevitably to learning.

From the behavioral point of view, the evolution from reception to learning appears inevitable. Reception is progressively aided by the development of mechanisms of sensory search, fixation, and attention. To be efficient, reception involves both differentiation and generalization. Generalization merges into transportation and transfer, and at this point any sharp separation of unlearned and learned functions ceases. The development of a maximally effective receptor system leads to the formation of mechanisms and processes basic to learning or involving learning which directly improve the efficiency of operation of the receptor processes. They are the kinds of mechanisms and traits that should arise from multiple mutations, each mutation providing a slight increment of evolutionary gain.

Perhaps the most amazing example of convergent development is the evolution of color vision, which has been found in some fishes (Teleostei), some insects, some birds, and most primates. As far as we know, primates are the only terrestrial animals with color vision. In spite of the diversity of forms in which it has evolved, color vision is remarkably similar in all. If one plots a curve showing difference thresholds as a function of wave length, the curves for pigeon and man are almost superimposable. There are no data to suggest that the color vision of man and fish differs in any radical way; and the greatest deviant, the bee, differs primarily in the ability to see into the ultraviolet. It is puzzling, although of course no evolutionary problem, that the primates are the only mammals with hue discrimination. Even more puzzling is the problem of the cat. Granit (1955) has shown that the cat has the retinal structures and functions always associated with color vision, and Lennox (1956) has demonstrated that the cat has, in its lateral geniculate body, the structures always associated with color vision. But the cat is totally color blind.

The existence of color vision in radically different forms of animals, and even radically different forms of eyes, must illustrate the adaptive value of a relatively slight modification in receptive capability, at least in the field of visual reception. Color vision must have evolved slowly and must have operated within the principle of orthoselection. There is very scanty evidence from primate data that color vision first consisted of differential response by lemurs (presumably archaic lemuroid forms) in the blue region of the spectrum (Biernes de Haan & Frima, 1930). It is a matter of established fact that, as we go from catarrhine monkey to anthropoid ape to man, the ability to see in the long-wave portion of the spectrum progressively improves (Grether, 1939, 1940). Indeed, if we exclude the spider monkey, learning ability and hue discriminability within the primate order would be almost as highly correlated as learning ability and complexity of cortical structure.

Comparing the learning capacities of fish (Reeves, 1919), honey-bees (Frisch, 1914), pigeons (Hamilton & Coleman, 1933), and primates shows clearly that the evolution of color vision does not of necessity imply an equal level of learning capability. Whether or not there is a high correlation between evolution of color vision and learning ability within a particular class of animals cannot be resolved in terms of the extant data. Color vision has been demonstrated in teleost fishes but has not been tested in any other superorder of fishes. The insects known to have color vision, the bees, have been shown to possess startlingly complex behavior patterns, whether or not these are subject to learned modifications. Recent studies have shown that the pigeon has a wider range of behavioral capacities (Ginsburg, 1957; Reeves, 1919) than had previously been believed, and it is more than possible that the pigeon and other birds are more capable of solving moderately difficult learning problems than many mammals. Finally, the primates as an order are preeminent over all other mammals both in terms of their visual capabilities and their capacity for learning.

The assumption that the evolution of learning was dependent in large part upon the evolution of receptors in no way precludes the likelihood that increased learning capability frequently operated as a selective factor during evolutionary development. There are, however, reasons to believe that other factors of great, and perhaps predominant, importance were involved. The process that began to separate the primate order from the other mammalian orders was not an increased capacity but the development of the visual mechanisms, which apparently arose as the pre-primate ancestral forms adapted to an arboreal life, an adaptation which would provide considerable environmental isolation. It is a striking fact that the brain of the primitive mouse lemur, *Microcebus*, shows the tripartite calcarine fissure and a lateral geniculate body with six distinct layers (Clark, 1934). From our scanty knowledge of more advanced lemurs, we may presume that *Microcebus* would show no unusual learning capacities, but the matter remains to be decided by direct test. The brain of *Tarsius* also shows striking development of the visual mechanisms, which, observational evidence suggests, arose independently of any striking gain in learning capacity. Thus, our information about primitive forms favors the view that the remarkably advanced visual system of the primates antedated their preeminent learning capacities. It is certain that man's primitive ancestors showed no less early complex development within the visual system, and it is certainly not chance that such a receptive system was antecedent to the development of the human visual cortex.

According to present-day theory, evolution takes place by natural selection among multiple mutations, each of which produces some small organismic change. Such a position appears to be at variance with the evolution of learning, if we think of learning in terms of our everyday terminology, for the language of learning implies the appearance of rather radical changes in capabilities. The Platonic specter of the national mind is still among us, with the implication that there is some broad gulf separating human and subhuman learning, or at the very least that some set of cumulative changes arose with startling rapidity as man diverged from other higher primate forms.

If we are to explain learning in terms of modern evolutionary theory, there should be continuity from the simplest to the most complex forms of learning. The appearance of a radically new kind of learning at any evolutionary point or period, including that during which man developed, is not in keeping with modern gene theory. Yet we find such an eminent authority as Dobzhansky writing, "Man is not simply a very clever ape, but a possessor of mental abilities which occur in other animals only in most rudimentary forms, if at all" (Dobzhansky, 1955, p. 338).

Dobzhansky falls into the common error of assuming that the particular human traits of language and culture imply the existence of some vast intellectual gap between man and other animals. The probability that a relatively small intellectual gain by man over the anthropoid apes would make possible the development of symbolic language and also culture is given small consideration. It is a common error to fail to differentiate between capability and achievement. Thus, the fledgling swallow, a few days before it can fly, differs little in anatomical and physiological capacity from the swallow capable of sustained flight, but from the point of view of achievement the two are separated by what appears to be an abysmal gulf.

By comparing selected traits about which we are ignorant, rather than those about which we are informed, one can argue for great intellectual differences between ape and man. We have little knowledge concerning the "aphasia" characterizing monkey and ape, but such little knowledge as we have suggests that the anthropoid ape's language inadequacies basically result from the failure to possess certain unlearned responses (Hayes, 1951). The degree to which it is additionally dependent upon intellectual differences is unknown. The failure of chimpanzees to develop culture may have resulted from some small but critical deficiency in intellectual ability or in specialized unlearned responses, such as those underlying tool construction or nonemotional vocal evocation. Other nonintellectual factors preventing cultural development may be the lack of social groups of an essential critical size, or limitations imposed by the physical environment. Knowing nothing about these factors permits unlimited speculation on the part of the scientist wearied by the research routines required by his field of specialization.

In contrast with these areas of ignorance there has gradually developed during the last quarter of a century a rich experimental literature comparing intellectual performances among many species within the primate order. No one suggests that more than a beginning has been made, but such data as are extant question the assumption that there is a wide intellectual separation between the human and the subhuman primates. The explanation and ordering of these data may be done best in terms of a simple classificatory schema.

Many learning problems can be classified effectively in terms of the complexity of the factors which interfere with successful problem solution. A well-studied learning task of moderate difficulty is the object or cue discrimination illustrated in Figure 17. Two stimuli, such as a triangle and a circle, are placed over the two foodwells of a test tray. One stimulus, the triangle, is consistently rewarded, although it is on the right side of the tray on Trial 1 and on the left side on Trial 2. The position of the triangle varies in an irregular but balanced manner during the learning trials.

Now it is obvious that on a single correct trial both the triangle and the position it occupies are simultaneously rewarded, and because a particular position as well as a particular object are rewarded, the nature of the reward is ambiguous rather than differential. During the many trials, however, the triangle is rewarded on every trial, and each of the two positions, right and left, is rewarded on only half the trials. The inconsistent reward of the ambiguous position cues apparently leads to their elimination, and learning the

Fig. 17. Object discrimination with a triple-prong and a cone.

object-discrimination problem may be described as the inhibition or suppression of the positional response tendencies.

From the point of view of the complexity of the ambiguity of cues, the object-discrimination problem is relatively simple. Only one condition of ambiguity of reward exists, ambiguity between the object and the position rewarded. This comparatively easy problem can be solved by a wide range of animals, including fish, mice, rates, pigeons, cats, and dogs, as well as monkeys, apes, and men.

The oddity problem involves a single additional condition of ambiguity, and the problem is illustrated in Figure 6. Three stimuli, such as two circles and a triangle, or two triangles and a circle, are placed over the three foodwells of a test tray. The odd, or singly represented object, is correct, and the other two objects are incorrect regardless of the position they occupy and regardless of the correctness or incorrectness of any object on any previous trial. Thus, as is illustrated, the triangle may be correct on Trial 1 and the circle on Trial 2. Now it is obvious that on any particular trial both the position rewarded and the object rewarded are ambiguous. Thus, there are two ambiguous factors in contrast with the single ambiguous condition found in the object discrimination problem. It is hard to believe that solution of a problem involving the addition of a single error factor requires some new "mental ability." From the standpoint of achievement, however, a sudden separation has appeared between various orders. No pigeon, rat, cat, or dog has solved the illustrated oddity problem, even though pigeons (Ginsburg, 1957) rats (Wodinsky & Bitterman, 1953), and cats (Warren, personal communication) are reported to have solved simplified versions of this problem or problems of generally similar type. Furthermore, we are now dealing with a task that is beyond the intellectual capacity of the young child, although data defining the minimal human chronological age level for oddity problem solution are lacking.

A third condition of ambiguity is added in the combined oddity-nonoddity problem. The stimuli are the same as those used in the oddity problem, but the odd or the nonodd object is correct, depending upon the color of the test tray. A green test tray may indicate that the odd object is correct, an orange test tray that the nonodd objects are correct. On any particular trial there are three factors which are ambiguously rewarded—position (right or left), object (triangle or circle), and configuration (oddity or nonoddity). Again, the introduction of an additional condition of ambiguous reward has greatly increased problem difficulty. No subprimate animal has as yet been reported to approach solution of this problem, which can be mastered by monkeys and apes without undue difficulty. Problems of this type can be recognized as components of human mental tests. Although there are no definitive data on the chronological age at which these triple-ambiguity problems can be solved by the child, it is certain that a vast number of human beings, including adults, cannot master problems of this class of difficulty.

Even more complicated problems can, however, be solved by monkeys (Spaet & Harlow, 1943; Young & Harlow, 1943) and apes (Nissen, Blum, & Blum, 1948). One chimpanzee solved tasks involving five conditions of reward ambiguity (Nissen, 1951a), and the color-categorizing performance attained by one rhesus subject (Weinstein, 1955) was disarmingly humanoid. Effective reviews of the performances by monkeys on complex learning tasks are available in the psychological literature (Harlow, 1951a, 1951b; Nissen, 1951a).

The particular kinds of problems chosen and the classification selected were not capricious. Problems of the kinds described have been used to measure human conceptual abilities (Goldstein & Scheerer, 1941; Roberts, 1933) and to differentiate normal and brain-injured patients (Weigl, 1941). We have not taken a particular type of test peculiarly adapted to the monkey and used it to test man; we have taken a particular type of test peculiarly adapted to man and used it to test monkeys and apes. If there is any improper comparison in the use of the described tests, it must be unfair to the subhuman, not the human, animal. Be this as it may, the tests clearly demonstrate that defining man as "a possessor of mental abilities which occur in other animals only in most rudimentary forms, if at all" (Dobzhansky, 1955) must, of necessity, disenfranchise many millions of United States citizens from the society of Homo sapiens.

If one appraises factually and unemotionally the learning data of animals on problems ranging in difficulty from object discrimination to effective measures of human conceptualization capabilities, one cannot help but be struck by the intellectual kinship among the phyletic groups being tested. There is no evidence of an intellectual gulf at any point, and there are no existing data that would justify the assumption that there is a greater gap between men and monkeys than there is between monkeys and their closest kin below them on the phylogenetic scale.

As learning and so-called thinking problems become more complex, the number of ambiguities among the problem components increases, and the possible number of extraneous and inappropriate responses increases. All learning and all thinking may be regarded as resulting from a single fundamental operation, the inhibition of inappropriate responses or response tendencies. Since such a position may seem radical, we briefly review alternative positions taken by modern learning theorists.

All learning theorists can be divided into two groups. One group assumes that learning is the resultant of two opposed mechanisms, which are usually described as excitatory and inhibitory mechanisms, the former strengthening earlier responses or response tendencies and the latter weakening such responses and response tendencies. The advocates of such a theory may be described as duoprocess learning theorists. A converse position asserts that all learning is the result of a single process, either an excitatory or an inhibitory process. Advocates of such a position would be called uniprocess learning theorists. Hebb (1949), for example, appears to be a uniprocess learning theorist favoring an exictatory process.

I have recently presented experimental and theoretical evidence favoring the uniprocess position (Harlow & Hicks, 1957), but, contrary to Hebb, I believe that inhibition is the single process accounting for all learning. It is presumed that this unitary inhibitory process acts to suppress the inappropriate responses and response tendencies operating to produce error in the problems just described. The nature of inappropriate response tendencies found in the object-discrimination (Harlow, 1950) and oddity problems (Moon & Harlow, 1955) has been analyzed in detail. Although I have not yet presented the position formally, I have believed for a number of years that the development of all complex learning of the type described could best be explained in terms of uniprocess inhibition theory.

Having come to this opinion, the data on learning by primitive organisms presented for me a new and intriguing problem—investigation of the possibility that the simplest as well as the most complex learning tasks might fit into uniprocess, inhibitory learning theory, that simple as well as complex learning problems might be arranged into an orderly classification in terms of difficulty, and that the capabilities of animals on these tasks would correspond roughly to their assigned positions on the phylogenetic scale. If true, the combined data would present the possibility of ordering all learning phenomena from habituation to abstract thought within a single system in which all differences would be explainable in quantitative terms.

A number of investigators have reported learning in the paramecium; the investigation by French (1940) is typical of the type of function measured, and is the best controlled of the studies. French sucked paramecia into a glass tube .6 mm. in internal diameter and recorded the time required for the animal to swim from the lower end of the tube into its individual culture medium. Ten of the 20 paramecia used as subjects showed significant decrease in escape latency. French observed that during the first few trials these animals would swim back and forth only a few times and then make one long dive to the bottom. These data can be interpreted in terms of learning to inhibit responses extraneous to the culminating response of culture-medium entry. His data show clear-cut inhibition and consequent reduction of activity, and give no evidence for the formation of new associations.

Fleure and Walton (1907) reported learning data for the sea anemone which are also in accordance with a learning theory that stresses inhibition-type learning. These experimenters placed pieces of moist filter paper at 24-hour intervals on the same groups of tentacles of several sea anemones. At first the paper was carried to the mouth by the tentacles, but after two to five repetitions of the stimulus it was rejected by the animals; this habit was retained for 6 to 10 days for the specific set of tentacles involved. The nonstimulated tentacles accepted the filter paper even while the trained tentacles rejected it. The only learning that appears to have taken place here is inhibition of the original responsiveness to filter paper. The persistence of the inhibition in a relatively simple organism is striking.

One of the most comprehensive and best controlled experiments on learning in a lower form is the study by Hovey (1929) on the marine flatworm, genus *Leptoplana*. Hovey produced conditioned inhibition of the photokinetic response in 17 flatworms by exposing the individual subjects to light for a series of 5-minute periods and inhibiting progression by tactual stimulation of the anterior tip of the worm whenever the animal began to creep in response to light stimulation. The mean number of tactile stimulations required to inhibit movement steadily declined from 110 on Trial 1 to 5 on Trial 25; and, even though none of the animals developed perfect inhibition of the photokinetic response, some subjects were completely nonresponsive during individual 5-minute test sessions. Control groups of subjects ruled out the possibility that the results could be explained in terms of fatigue, light adaptation, or injury to the snout.

It is perhaps no accident that the first kind of conditioned response to be described in a primitive animal is a conditioned inhibition of an innate, so-called tropistic response. The learning consists in learning to inhibit this response tendency.

Robinson (1953) measured the learning by earthworms, *Lumbricus terrestris*, of a single-unit T-maze, a spatial discrimination problem. Robinson's study is selected for presentation because, though the design and results are similar, it was conducted at a level of technical elegance surpassing the pioneering investigation by Yerkes, or the subsequent studies of Schwartz or Heck. He recorded time and errors and traced records of the earthworms's behavior. By preliminary tests, Robinson determined the preferred maze arm for each worm, and learning was measured in terms of reduction in frequency of entrances into the preferred arm. Partial entrance into this arm resulted in contact with an electrode and consequent electric shock; choice of the correct arm led to entry into a goal box which was also the subject's living box. Robinson's data make it quite clear that the earthworm did not develop any new association between turning into the incorrect alley and shock. Instead, the shock resulted in agitated and excess movements and persistent negative reactions to the stimuli on both the correct and incorrect sides of the junction. These negative reactions were gradually reduced—or, as we would say, inhibited—but the negative reactions even to the stimuli on the correct side of the junction were never entirely eliminated. Robinson's results are strikingly similar to those presented by Hovey despite marked differences in the test situations. There is no sudden emergence of any new association; there is only the gradual elimination or inhibition of the extraneous responses.

Discrimination learning has been reported by Boycott and Young (1950) in the octopus. A crab lowered into one end of a tank served as the positive stimulus, and a crab attached to an electrified white plate was used as a negative stimulus. Within 25 trials the octopus came to inhibit response to the negative stimulus in about 85% of the trials, whereas the only change to the positive stimulus was some slight inhibition, but in no case enough to prevent contact. Again, there is no evidence that learning resulted in the strengthening of any response. Indeed, the converse happened: The inappropriate response was almost completely inhibited, and even the appropriate response was partially inhibited.

If one surveys the literature on subvertebrate learning, a single fact stands out in a very unequivocal manner. Learning, all learning which has been adequately described and measured, appears to be the learned inhibition of responses and response tendencies which block the animal or fail to lead it to some terminal response, such as eating or escape from noxious stimulation. Furthermore, it should be remembered that many of these lower animals have reasonably well-developed, synaptic-type nervous systems, and there is no reason, from an anatomical point of view, to suspect that the nature of learning is going to be altered in any subtle, fundamental manner as we progress to higher forms. The law of parsimony requires, at the very least, that we seek as simple a fundamental explanation of vertebrate learning—including human learning—as is consonant with fact.

The concept of a kind of learning simpler than that of conditioning has been held by other investigators, including both Thorpe (1943) and Schneirla (1934). Schneirla, for example, describes "habituation learning," defined as learned conformity to a new situation through generalized adjustments involving inhibition of initial avoidance or shock reactions. This learning is differentiated from sensory adaptation. It should be kept in mind that the possibility definitely exists that no fundamental distinction can be made between sensory adaptation and learning, if we ever reach a level where these processes can be described in biochemical terms.

Ignoring, however, the relationship between sensory adaptation and learning, I wish to take the position here that there exists no fundamental difference, other than complexity, between the kind of learning listed as habituation learning and the kinds listed as conditioned response learning—and, for that matter, the kinds of learning described as reasoning and thinking.

I hold this position in spite of the fact that the Pavlovian-type conditioned response has been taken by many as the paradigm for all learning, with the assumption that a conditioned response involves the formation of some new association. This new connection is formed between the conditioned stimulus (CS) and the unconditioned stimulus (US) or between the CS and the UR (unconditioned response), depending upon the psychological learning school in which you happen to hold club membership. If a CS of light or sound is presented to an animal, followed by US of shock to the leg, the animal will—under certain circumstances—learn to flex the leg to the light or sound. Conditioned response formation is characteristically described as if a new connection had been formed. It is, however, obvious that the animal already possessed the potential capacity to flex the leg to either the light or sound. There is practically no external stimulus of moderate intensity that does not have the capability of evoking any and all the postural responses of the body, and this position receives strong support if one considers the ontogenetic development data. Both visual and auditory stimuli of low intensity can, under favorable conditions, even in the adult, elicit exploratory patterns involving all or any part of the body musculature. From this point of view, we may give serious consideration to the fact that conditioning does not produce new stimulus-response connections but that it operates instead to restrict, specify, and channelize stimulus-response potentialities already possessed by the organism. It is entirely possible that a specific visual conditioned stimulus comes to elicit a specific leg flexion response because the presentation of the CS and US in a specific temporal pattern produced inhibition of response to extraneous and distracting external stimuli and inhibition of the other postural responses which the CS was already capable of eliciting.

The description of the learning of the spatial discrimination problem in terms of a uniprocess inhibition theory presents no challenge. A typical spatial discrimination problem involves learning by the animal to choose the right or the left alley of a single-unit T-maze. The common procedure is to give the subject experience on a straightaway unit with food at the end, until the response of running until food is received is thoroughly established. When the animal is transferred to the spatial discrimination problem, this task is automatically learned if the animal inhibits any response tendencies to run down the unrewarded alley and inhibits tendencies to be distracted by extraneous external and internal stimuli.

From spatial discrimination we pass next to nonspatial discrimination learning, also known as object or cue discrimination learning. This kind of learning has been exhaustively tested in the rat with the Lashley jumping box and in the monkey with the Klüver-type test tray. The animal must learn to choose one object, which is consistently rewarded, and not to choose the second object, which is never rewarded. The reader will recognize this as the problem with which we began in the classification of problems in terms of complexity of factors interfering with learning.

Such evidence as exists suggests that the learning of the object discrimination problem may well be explained in terms of inhibition of extraneous and inappropriate responses. Detailed analysis of this problem enabled us to identify four reaction tendencies producing errors (Harlow, 1950), and as these reaction tendencies are inhibited, the percentage of correct responses increases progressively until it approaches 100.

We have been able to carry our analyses of the factors which produce errors to a problem of still greater complexity, the oddity learning problem, and we can see no fundamental theoretical difference between oddity learning and discrimination learning, other than the addition of one or possibly two new error factors (Moon & Harlow, 1955). The data offer no indication that learning here consists of anything other than the inhibition of response tendencies which interfere with the culminating response leading to reward. We are convinced that we can carry these analyses even further, and we would be very much surprised if there is any fundamental differences in the learning of the oddity problem and the learning of differential equations—other than that of complexity.

Learning by amoeba and learning by man appear superficially to be such basically separable processes that I initially avoided direct comparison. I first chose to select classes of problems illustrating similarities and differences in the abilities of primates and other animal forms known to be gifted in terms of their learning abilities. Then I described the learning abilities of animals, including primitive animals, on the simplest known learning problems. The fundamental similarity between these problems and problems of greater complexity was demonstrated, until we reached the class of problems initially presented here. In spite of the fragmentary nature of research in the evolution of learning, it became obvious that position of an animal in the phyletic scale is related to complexity of problems it is able to solve, and it may be stated that complexity of most of the described problems had been independently defined both on logical grounds and on the basis of ontogenetic development in the higher animals.

The existing scientific data indicate a greater degree of intellectual communality among the primates, and probably a greater communality among all animals, than has been commonly recognized. There is no scientific evidence of a break in learning capabilities between primate and nonprimate forms. Emergence from the ocean to the land produced no sudden expansion of learning ability. Indeed, there is no evidence that any sharp break ever appeared in the evolutionary development of the learning process.

That this is probably true should surprise no one. Indeed, the fundamental unity of learning and the continuity of its developing complexity throughout phylogenesis, or at the least within the development of many major branches of the evolutionary tree, would seem to be in keeping with modern genetic theory.

CHAPTER 4

LEARNING TO LEARN

In most psychological ivory towers there will be found an animal laboratory. The scientists who live there think of themselves as theoretical psychologists, since they obviously have no other rationalization to explain their extravagantly paid and idyllic sinecures. These theoretical psychologists have one great advantage over those psychological citizens who study men and women. The theoreticians can subject their subhuman animals, be they rats, dogs, or monkeys, to more rigorous controls than can ordinarily be exerted over human beings. The obligation of the theoretical psychologist is to discover general laws of behavior applicable to mice, monkeys, and men. In this obligation the theoretical psychologist has often failed. His deductions frequently have had no generality beyond the species which he has studied, and his laws have been so limited that attempts to apply them to man have resulted in confusion rather than clarification.

One limitation of many experiments on subhuman animals is the brief period of time the subjects have been studied. In the typical problem, 48 rats are arranged in groups to test the effect of three different intensities of stimulation operating in conjunction with two different motivation conditions upon the formation of an isolated conditioned response. A brilliant blitzkrieg research is effected—the controls are perfect, the results are important, and the rats are dead.

If this *do and die* technique were applied widely in investigations with human subjects, the results would be apalling. But of equal concern to the psychologist should be the fact that the derived general laws would be extremely limited in their application. There are experiments in which the use of naive subjects is justified, but the psychological compulsion to follow this design indicates that frequently the naive animals are to be found on both sides of the one-way vision screen.

The complexity of the mechanisms of the human nervous system involved in the learning process suggests the existence of some active, underlying, coordinating organization of principles. There is nothing passive about the transformation and development of the learning capacities from infancy to maturity. Early learning theories have ranged from a dependence upon the feeble fumbling of blind trial and error learning to the opposite extreme of a belief in innate insight through which relationships are revealed.

The variety of learning situations that play an important role in determining our basic personality characteristics and in changing some of us into thinking animals is repeated many times in similar form. The behavior of the human being is not to be understood in terms of the results of single learning situations but rather in terms of the changes which are effected through multiple, though comparable, learning problems. Our emotional, personal, and intellectual characteristics are not the mere algebraic summation of a near infinity of stimulus-response bonds. The learning of primary importance to the primates, at least, is the formation of learning sets; it is the learning how to learn efficiently in the situations the animal frequently encounters. This learning to learn transforms the organism from a creature that adapts to a changing environment by trial and error to one that adapts by seeming hypothesis and insight.

The rat psychologists have largely ignored this fundamental aspect of learning and, as a result, this theoretical domain remains a terra incognita. If learning sets are the mechanisms which, in part, transform the organism from a conditioned response robot to a reasonably rational creature, it may be thought that the mechanisms are too intangible for proper quantification. Any such presupposition is false. It is the purpose of this chapter to demonstrate the extremely orderly and quantifiable nature of the development of certain learning sets and, more broadly, to indicate the importance of learning sets to the development of intellectual organization and personality structure.

Learning sets measure generalization of learning. Instead of following the conventional model of training an animal on a single problem to a preestablished criterion, we trained monkeys for a limited number of trials on a single problem of one general class or type of problem and then, regardless of success or failure, we trained them on many different but similar problems of the same class or type. The first series of problems were all on object-quality discrimination learning. To refresh the memory on object-quality discrimination learning, the subject is rewarded for choosing one of a pair of objects which differ, for instance, in form, such as a triangle or a circle, or in color or in degree of brightness. The series included 344 such problems, 32 of which were preliminary discriminations. Each problem used a different pair of stimulus objects. Thanks to our previous studies, we knew that monkeys could learn discrimination problems rapidly.

On the first few problems presented, the learning process of the monkeys followed the fumbling of trial and error learning, but each new problem based on the same principle was mastered with increasing ease and efficiency. By the time the problem was solved on the very first trial, even the manner of the monkey's learning had changed dramatically. Forceful decision replaced fumbling as insight appeared, and, if the incorrect object were chosen on the first trial, there was an immediate shift to the correct choice and no further errors occurred. Choice of the correct object on the first trial practically precluded further errors.

Many learning theorists had delegated insight learning to be a divine right of the human primate while trial and error learning was the ultimate Thule of the subhuman animal. The learning set tests demonstrate that trial and error and insight learning are two different stages of one long and continuous learning process, part of the well-ordered development of learning.

We also have the results of a long series of discrimination problems presented to a different primate species. Nursery school children from 2 to 5 years of age were chosen as subjects because in young children a minimum of experience conceals and confuses the basic behavior. Brightly colored macaroni beads and toys replaced the food rewards and iron-barred cages were not required, in spite of the behavioral vagaries sometimes common with this particular primate species. Although as a group the children were more rapid learners than the monkeys, they, too, made many errors early in the tests and the learning set curves portray an orderly, gradual, progressive increase in the percent of correct responses. The children made the same type of errors before learning to solve the problem in one trial, and the brightest monkeys outperformed the slowest of the nursery school tots.

The formation of each learning set represents the acquisition of an organization of habits appropriate to solution of a certain type of problem. An iota of insight into the complexity of human learning and thinking is provided by reflection upon the infinite number of learning sets which enable the human being to adapt to the demands of everyday environment and of an everchanging environment. Subsequent tests given to both the monkeys and the young children reversed the object conditions of the discrimination test. The correct and incorrect choices were switched so that the previously incorrect object came to be rewarded. In this type of test, the differences between the two species were more marked and the children learned much more rapidly to shift the choice of object after only a single failure.

When we started to study learning set, there was no scientific background for the subject and no scientific models from which to work. Obviously, in studying learning set one wants to measure performance on a maximum number of problems. To prevent the experiment from extending to eternity, one must then make the decision on the number of trials per problem. The number of trials was by definition almost guesswork, and the part not guesswork has never been done.

The first trial of each problem is a blind trial and cannot be used as a measure of learning. If the monkey has a high stimulus preference for the object which proves to be correct on the first trial, it will commonly solve the problem immediately without any errors. This gives you a phenomenon called insightless insight. If, however, he first chooses a nonpreferred object which proves to be correct, a large number of further trials will be required in overcoming this negative preference choice. It becomes very obvious that a single problem, as well as a single trial, is a very ineffective measure of any learning at all unless you have absolute control of the monkey's preferences. This can be achieved by asking the monkey, but he never answers.

The second trial of any problem measures the amount learned on the preceding trial. If the amount is little, as in learning by trial and error, the percentage correct will be low. However, if the learning set has been perfectly mastered, the learning set from Trial 1 will insure perfect mastery on Trial 2, i.e., 100% correct responses on all problems of a series. One way to measure increasing proficiency during the course of learning set formation is to take the percentage of correct responses on Trial 2. It is unbiased by any other factors, by any factor in intraproblem learning. Percentages of correct responses on Trial 2 during learning set formation of rhesus monkeys will average 70% correct for the first hundred problems, 80% correct for the second hundred problems and 90% correct for the third hundred problems.

There are a host of interesting problems that relate to the number of trials that each problem in a learning set test should have. There is some reason to believe that the early problem should be long so that even the stupid can get some gain. The number of trials per problem would then be reduced as the animal mastered the learning set principle. Theoretically, in learning set formation, the animal must obtain some information from the early problems and since the early problems are blind, it follows that early problems should have many trials. As he learns faster and faster, the number of trials in each problem should be reduced. Again, theoretically the number of trials per

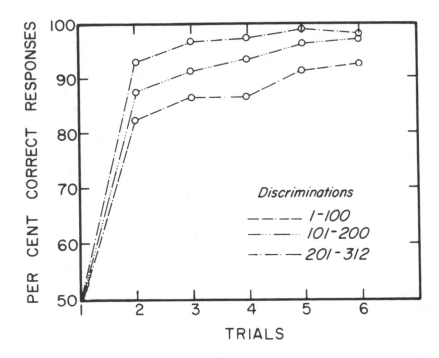

Fig. 18. The rapid improvement in correct responses on and after Trial 2, until learning set is almost perfect.

problem should be high if one tests stupid animals, such as stupid children. In reality, with stupid animals, such as guinea pigs and rats, the number of trials in the early problems should be infinite, since learning will be little and transfer from problem to problem will be essentially nonexistent. The rat will give everything to the problem, but he has nothing to give. Since the number of trials per problem was guesswork and since it is a variable of not overwhelming significance, within certain limits, we made the arbitrary decision to run each of the discrimination problems in a series of similar problems for 6 trials and then to go on to another problem. Levin shows that if the problems are only 3 trials in length, the monkey would facilely form the learning sets. Subsequently Bowman showed that learning set formation was entirely possible when each problem was only 2 trials in length. I have always been convinced that if I had had more graduate students, we would have found that learning sets were possible if we had no trials at all. The practical solution is to run each problem for 6 trials. This is a conservative procedure and a large number of problems can be run every day.

There may be some gain from measuring learning set formation in terms of performance on Trial 2, since this is measuring the degree to which problems are being changed from trial and error to insight learning. Trial 2 is an insight measure. However, all choice of scores is arbitrary and a measure which we have often used is performance on Trials 2 through 6 combined. Of course, we never measured Trials 2 through 6 when measuring pretrial learning set problems, but no doubt somebody will.

Figure 18 presents learning curves showing the percentage of correct responses for the first 6 trials of this entire final series of 312 problems on object-quality discrimination. The monkeys learn how to learn individual types of problems with a minimum of errors. This learning how to learn a certain kind of problem is what we designate by the term learning set. The three curves of the final 312 problems show the rapid improvement in percentage of correct responses as the learning set becomes established, until an almost perfect percentage correct is maintained after Trial 2 during the last problems.

We wish to emphasize that this learning to learn, this transfer from problem to problem which we call the formation of a learning set, is a highly predictable, orderly process which can be demonstrated as long as controls are maintained over the subjects' experience and the difficulty of the problem. Our subjects, when they started these researches, had no previous laboratory learning experience. Their entire discrmination learning set history was obtained in this study. The stimulus pairs were equated for difficulty. It is unlikely that any group of problems differed significantly in intrinsic difficulty from any other group.

In a conventional learning curve we plot change of performance over a series of trials; in a learning set curve we plot changes in performance over a series of problems. It is important to remember that we measure learning set in terms of problems, just as we measure habit in terms of trials.

For the 344 problems presented to the monkeys, Figure 19 plots the learning curve on the basis of Trial 2 response only. This curve shows dramatically the effectiveness of the first trial, or, in other words, what happens when the learning set is really taking hold and insightful learning occurs. After the formation of the discrimination learning set, a single training trial constitutes problem solution. The learning set fulfills its function by transforming a problem from a tribulation to a triviality. Lest the one trial solution give a false impression, however, it should be remembered that the one trial learning appears only after an orderly learning process with the preceding problems of similar nature.

When the same rhesus were tested on discrimination reversal problems, we had a change in the number of discrimination trials preceding the reversal phase, so that the animals could not learn that the problem of going to shift after any fixed number of trials. The discrimination phase was run for either

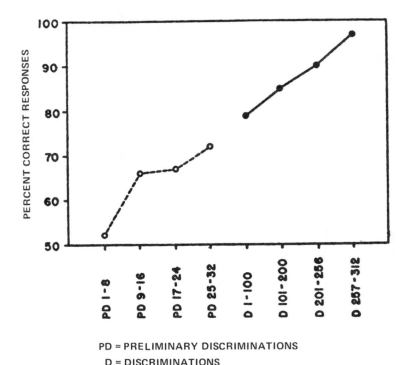

PD = PRELIMINARY DISCRIMINATIONS
D = DISCRIMINATIONS

Fig. 19. Insight learning: Discrimination learning set curve based on Trial 2 responses.

7, 9, or 11 trials and at that point the stimulus that was previously incorrect became correct. It was obvious that many of the monkeys tried very, very hard to guess. They knew that the experimenter was going to pull a fast one from his bag of tricks, but they could not find out just when.

One group of monkeys received a special bonus and was trained on a final refinement of the discrimination-reversal problem. The object rewarded was shifted for one trial only and then another shift occurred, back to the reward of the original object. After confusion during a number of problems, the monkeys completely ignored the one trial shift and it was just as if they were politely pretending that the experimenter had made an error, but they would overlook the mistake.

The discrimination reversal problems might have been expected to cause the monkeys more trouble than the original discrimination series, and had they been presented first this might well have been true. As it was, they proved easier. The discrimination reversal set was formed more rapidly. A

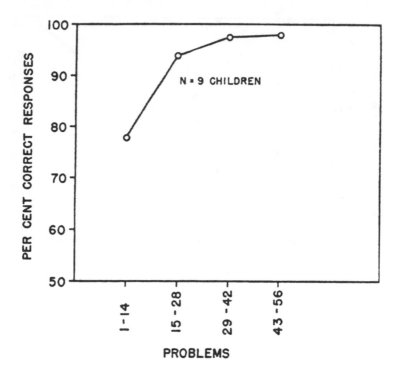

Fig. 20. Percent of correct Reversal Trial 2 responses by children on 14-problem blocks.

detailed analysis of the second learning process indicated that there was transfer of training from the first series of problems which may, to oversimplify, be called transfer of a set to form a set. The monkey acquired a generalized ability to learn any discrimination problem or any discrimination reversal problem more easily. Training did not turn the animal into a stereotyped automation but, instead, increased his capacity to adapt.

Before leaving the problem of discrimination reversal learning, we submit one additional set of data that we feel merits attention. Nine of the children previously referred to were also subjected to a series of discrimination reversal problems. The outcome is partially indicated in Figure 20, which shows the percent of correct Reversal Trial 2 responses made on successive blocks of 14 problems. It can be seen that these 3- to 5-year-old children clearly bested the monkeys in performance on this series of problems. Trial 2 responses approach perfection in the second block of 14 discrimination reversal problems. Actually, over half of the total Trial 2 errors were made by one child. These discrimination reversal data on the children are the perfect illustration of set formation and transfer producing adaptable abilities rather than specific bonds. Without benefit of the monkey's discrimination reversal set learning curves, we might be tempted to assume that the children's data indicate a gulf between human and subhuman learning. But the extremely rapid learning on the part of the children is not unlike the rapid learning on the part of the monkeys, and analysis of the error-producing factors shows the same basic mechanisms are operating in both species.

Unlike the evanescent learning of discrete isolated problems or facts, the learning set, once mastered, is retained over long periods of time. One of the striking aspects of the monkeys' acquisition of learning sets was that they remembered them enough for use upon demand after long time intervals. Even after more than a year, a few hours of practice would restore levels of efficiency which took the monkey many weeks to develop.

The emphasis throughout this chapter has been on the role of the historical or experience variable in learning behavior—the forgotten variable in current learning theory and research. Hull's neobehaviorists have constantly emphasized the necessity for an historical approach to learning, yet they have not exploited it fully. Their experimental manipulation of the experience variable has been largely limited to the development of isolated habits and their generalization. Their failure to find the phenomenon of discontinuity in learning may stem from their study of individual, as opposed to repetitive, learning situations.

The field theorists, unlike the neobehaviorists, stressed insight and hypothesis in their description of learning. These theorists conveyed the impression that insight is a property of the innate organization of the individual. If such phenomena appear independently and not as a part of a gradual learning history, we have not found them in the primate order.

The cornerstone of the theory that animals possess some sort of innate insight was the work of the famous Gestalt psychologist Wolfgang Köhler (1925) in his studies of chimpanzees. These prime examples of the great apes used sticks to obtain bananas they could not reach themselves. With the sticks they knocked the bananas down and with the sticks they raked them in. They climbed the sticks or pole vaulted with them, assembled short sticks into longer and used them to reach higher than stacked boxes alone would allow. No animal other than man has yet stacked more than four boxes. Because the chimpanzee often made sudden solutions of these problems, Köhler interpreted these flashes of insight as evidence of an innate ability independent of learning. He was impressed also by the coherence of the action sequences preceding solution and by the change in the chimp's facial expression upon success.

Unfortunately, the early learning experiences of Köhler's chimpanzees when in the jungle were unknown to man and the later laboratory

Figs. 21 and 22. Tool using by Wisconsin Laboratory cebus monkeys.

experiences were not consistently or completely recorded. More recent studies at the Yerkes Orange Park Laboratory emphasize the role of gradual learning in tool using. The 4-year-old chimpanzees studied by Birch at first showed little ability to use sticks as tools, but over a period of days gradually learned to touch objects out of reach. After mastering simple problems, they still had difficulty with more complex tasks. In the footsteps of Birch, Schiller found that the younger the chimpanzee the more slowly it succeeded with a series of stick problems. In a group from 2 to 8 years, some of the young subjects required hundreds of trials to solve the simplest tasks, while old and experienced animals breezed through them with little practice. None of the apes produced sudden insight solutions.

Klüver (1933) and others have found evidence that some cebus monkeys may solve tool problems as complex as those solved by the apes. Figures 21 and 22 show typical tool-solving by a Wisconsin Laboratory cebus monkey. The pole-climbing seemed insightfully sudden, but, with the law of gravity still on the legal books how could it seem otherwise?

There has been a long controversy over the use of tools as weapons. Our tool-using cebus, who showed insightful use of sticks a number of times, used them as weapons at least once. When he and a friend were backed into the corner of a cage surrounded by five larger rhesus monkeys, our tool-using terror picked up a stick and struck, hitting at least one of his attackers. Although his stick defenses were not perfected, he at least kept his opponents at bay and the fight ended a draw with no one quartered. Although my co-observer was a minister of the gospel. I did not publish these observations for 20 years, since not all aspects of the truth are consistently regarded as science.

Data on human children lend support to the position that children as well as subhuman primates need practice prior to the solution of tool problems. Nursery school children tested by Alpert failed in five experimental sessions to solve the easiest test of Köhler simple chimpanzee problems.

Whenever the animal's entire learning history is known, whether that animal is monkey, ape, or child, that animal at first solves problems by trial and error. If the animal is of the human primate species, the learning history so quickly becomes complex and its origins obscured that the path of learning development is much more difficult to trace, but we know that only gradually does it pave the way for seemingly immediate, insightful solutions.

There is little doubt that the learning set is an organizing mechanism or principle which plays an important role in learning and in the development of complex thought processes. By the time a young adult has completed years of education, both formal and informal, she has a wealth, both in number and in kind, of learning sets from which to select. Her basic working and thinking material bears scant relationship to the unlearned responses and just simple, learned habits of the neonate.

Our studies in learning sets offer more suggestions for educators than just the building from simple to more complex patterns of learning sets as increasingly more difficult and abstract problems are met. We know that the rhesus monkey was successfully taught to master a variety of learning sets which varied from simple to complex and we know that conditions must basically be met, aside from the physiological learning capacity of the animal. For each species to be taught, the number of problems necessary to teach each learning set must be determined. These problems must all center around the same principle and only one principle at a time. The quality or characteristic varied from problem to problem must clearly express the principle concerned.

As learning and thought become progressively more involved, the teaching of a new learning set may often make use of previously learned sets. If these previously learned sets have been inadequately formed through any defect in the process, the new learning will be ineffective and even disruptive. Arithmetic or mathematics well illustrates the progressive importance of each learning set. When clearly formed, the learning set for fractions contributes to the formation of the learning set for decimals based on fractions of a total of 100. These, in turn, aid in the formation of the learning set for percents, also based on parts or more than parts of 100. However, neglect any fraction or percent of the process in the formation of these learning sets and the fractions become fractious and the teaching is criticized. This student later in life cannot balance either the budget or her bank book and the percentages of chance of a happy marriage is rapidly reduced.

The language of words produces symbols or signs which represent or call forth appropriate learning sets. In reality, the word is, in itself, a condensed learning set. When you cannot think of a person's name, you seek for other clues to produce the correct learning set. If you cannot make a decision, you may ask a friend for advice, kindly words to produce the pertinent learning set or solution to the problem.

Monkeys make similar use of signs and symbols when they form learning sets for oddity problems according to the color of a test tray. Subhuman primates had been the subjects of successful research for many, many moons, and during many of these moons, experimenters had been trying to teach them human speech. Chimpanzees received most of this vocal instruction until some bright psychologists with the Bell Laboratories discovered that the subhuman primates lack the proper anatomical structures to form human sounds of speech. About the same time, two other very bright psychologists, the Gardners, adopted Washoe, a fine girl baby chimpanzee and, most successfully, taught her the American sign language of the deaf. Gestures of all varieties come naturally to the subhuman primate, and Washoe's gestural symbols for learning sets came to be counted in the hundreds.

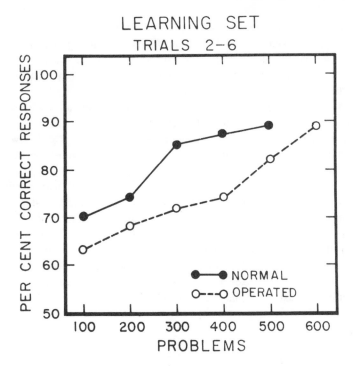

Fig. 23. Learning sets used in assessing cortical injury and damage.

The learning set principle has been used to gain information on a wide range of problems. It is a fairly effective measure of the maturation of learning in monkeys or of the learning maturity of the monkey. Individual problems can be solved earlier in the development of these animals than can the learning set principle. Learning sets have also been a useful tool for assessing cortical injury and cortical damages as shown in Figure 23. Actually, the discrimination learning set is an efficient technique for measuring interference or destruction in temporal lobe associative areas.

Hodos (1970) has collected data from a large number of sources in order to compare learning set performance on as many different species as directly comparable data could be obtained. The data was reported in the form of percent correct on Trial 2 during successive blocks of every 50 or 100 problems. The ability to discriminate is common to a wide variety of species and, by measuring improvement rather than absolute levels of performance, learning set gives a more comparable base of comparison than many other measures.

An anthropologist, Gregory Bateson, also described the principle of learning set or learning to learn under the name of deuterolearning. Bateson did not, however, formalize his construct in terms of any planned experimental procedures that would translate his hypothesis for use in further helping learning landmarks.

We ran a series of studies combining our learning set studies with our brain function studies, as far as our data permitted. It became perfectly obvious that our monkeys had formed adequate learning sets. They could withstand the insult of extensive brain injuries with impunity, or almost impunity. The learning sets remained.

More seriously, these data may indicate why educated people show less deterioration with advancing age than do uneducated individuals. The data lend support to the fact that our fields of greatest efficiency are the last to suffer gross deterioration. Learning sets coordinate facts and concepts, and principles do not fall as fast as facts. The only problem is to pick the learning sets that you want after half of the cortex is gone. However, the educated man can face arteriosclerosis with confidence, since the results on our brain injured animals must be applicable to man.*

*For further detailed data the reader is referred to Harry, H. F., The formation of learning sets. *Psychological Review*, 1949, **56**, 51–65. Copyright 1949 by the *American Psychological Association. Reprinted by permission.*

PART II

THE MEANING OF MOTIVES

CHAPTER 5

THE MATRIX OF MOTIVES

In our early learning studies we were not just looking around for new tests for their own sake or for some unknown future at stake. The original motivation was a search for problems definitive in establishing cortical localization of function. The human data which I knew from teaching physiological psychology showed that somewhere within the posterior association areas thinking was created and man was accomplished.

In the 1930s there was intense interest in proving or disproving Lashley's (1929) theory of equipotentiality of mass cortical action. Jacobsen's frontal lobectomies had ostensibly shown the intellectual specificity of the frontal lobes and our research on the occipital lobes was in accord with Jacobsen (1936). The 1930s, to be understood, were also the pre-electrical era or, as I prefer to call it, the pre-electrical error. After God created 12 billion neurons, why man should want to stimulate only one of them I could never understand. Instead of electrical stimulation, we cut. Had modern electrical techniques existed, I would still have relied on lesions. It was necessary to establish gross localization of function before even considering the single cell. Later we also aspirated. Freud already knew that sucking was of vital importance.

As the Primate Laboratory evolved, so did my thoughts. A new development emerged. My interest in cortical localization waned, although it never completely disappeared. My interest in learning as a broad, stimulating area of research took over as my primary consideration for many years. My learning interest, in keeping with evolution, bifurcated.

The bifurcation spread and broadened my behavioral concepts. I became interested in motivation. Learning psychologists had earlier used motivation to manipulate learning. To the contrary, I reached the point where I used learning to learn about motivation. Through the exigencies of the Primate Laboratory, I found that hunger, all by itself, was a variable of limited importance. If you wish to tame a monkey hunger may facilitate the elimination of fear, but during our learning experiments we had discovered that monkeys could not solve complex learning problems unless they were completely secure and happy in the situation. A hungry monkey is not a completely happy monkey.

Security was first achieved by patiently adapting the animals to the test situation. We reversed the hungry rhesus situation by taming them with small morsels of decent delicacies such as bananas and oranges. We waited and watched until the monkeys came to us for these small particles of food and then encouraged them to sit on our laps while feeding. We did not punish them, because pain breeds distrust whereas patient pleasure produces security and trust.

Hunger models are rich for rodents but maudlin for monkeys—and men. If early psychologists had picked an animal to be as different from man as possible, they would have picked the rat—and they did. There are many ways to do inadequate research, and this was an enormously insightful way.

Motives are commonly classified into two broad groups, the internal and the external motives, the motivational agencies inside the body and the material forces or incentives outside the body. Incentives pull the participant toward the gates and goals that lie without. They guide behavior toward external exigencies such as tastes, tests, facts, and even more important behavior such as fun and friendships.

When our learning research started it was dogma that learning depended upon innate homeostatic drives such as hunger, thirst, and sex. The internal motives were described and dignified by Cannon at Harvard University. They were potently publicized by Cannon (1929) who was a big gun in physiology and occasionally a big bore. Cannon produced and publicized the internal changes accompanying hunger, fear, and anger. Because of his enormous scientific prestige, the Cannon theory became widely accepted and the internal drives have commonly been named the primary motives.

One of the great positive contributions of Cannon's theory was the discovery of the hypothalamus. For decades, the hypothalamus did nothing other than serve as the site of some nasty neurological disorders. The hypothalamus became socially acceptable by assigning to it basic behavioral functions including hunger, thirst, satiation, and starvation, sex, and happiness. In view of the minuscule size of the hypothalamus and the enormous range of functions it subserves, never before have so few neurons made so many contributions to puritanical ethics.

It seems unlikely that the hypothalamus really mediated pleasure, since Sir Henry Head had previously assigned this triumphant trait to the thalamus itself. Sir Henry Head was always ahead of his field and he could not possibly have been wrong. Head was always ahead of his colleagues, and Head was never wrong.

There is general scientific sanction of the theory of homeostatis which supports the notion that the body initiates changes to restore optimal balance when optimal balance is upset. Richter (1942–43), as well as Cannon, created the myth, brilliantly conceived and based on homeostasis, that basic, primary motivation comes only from visceral stimulation inside the animal body.

Subsequent supporters have raised the internal drives to such a pedestal that motivation is mortally mutilated if the body is not in a deprived drive state. Many psychologists were overwhelmed by the elegant objectivity of Cannon's belligerent broadsides and internal drives became the brainless barn for motivational theories. Very soon, all learning became derived from changes accompanying the emotions and the hungers for food and thirst quenchers, to which sex soon became added. Thus, the knowledge for thirst became more important than the thirst for knowledge, and homeostasis, instead of being an emergency reserve, became the reason for the being of all behavior.

The peerless paragons of derived drives soon had its two greatest advocates, Hull (1952) and Spence (1950). According to these eminent authorities, all other human motives were derived, conditioned, and learned from the primary internal drives. Within a few decades, our thinking might have been totally dependent upon visceral agony and effort.

Our motivating position is that of the primacy of incentives, not the primacy of internal motives. No one underestimates the importance of the internal, primary drive motivating system. The external motives, the incentives do interact, but the relationship between the two systems changes considerably the higher up one goes on the phyletic scale. With the fish or the fowl and possibly the rodent, there is less interrelationship between the incentive system and the internal drives.

Incentives are not to be confused with fixed action patterns which are *responses* to external stimuli and are enthusiastically endorsed by ethologists. Fixed action patterns are released by very precise stimuli which, in turn, release very rigid behavioral responses. The spot on the gull's beak releases food ingestion in the young gull. The red color of the stickleback releases aggressive behavior in the other stickleback. Such releasing stimuli as these in the subprimate forms prevent and override any complex learned behaviors.

In the monkey, complex behaviors such as curiosity and exploration override any fixed action patterns. The importance of the primary, internal drives in man and monkeys is basically what they do to the incentives. Simple, unlearned behaviors produce simple unlearned responses, whereas complex unlearned behaviors, such as one finds primarily in monkeys and in man, produce the complex unlearned behaviors (Harlow & Mears, 1978) which lend themselves to learning (Figs. 24 and 25).

Many investigators found that with the subprimates, including the rat, learning was enhanced by length of deprivation of the same drive satisfier as was used as an incentive in the learning problem. If water was the incentive, rats

Fig. 24. A young model of complex curiosity behavior.

Fig. 25. Rhesus reproduction of the human model.

learned a maze more rapidly when water deprived. Rats increased their speed, their frequency of response in a Skinner box, faster after one day of deprivation of food than after one hour without food.

Even with monkeys the *rate* of pressing a lever in a Skinner-type box was increased more after 47 hours of food deprivation than after 25 to 1.

A brilliant graduate student who believed that monkey learning was related to hunger ran three groups of subjects on three stages of food deprivation: 0, 12, and 24 hours of food denial. He tested them on discrimination learning set formation, one of the more complex learning problems. This graduate student had read Hull. He discovered the Hull truth, but could not believe the whole truth. There was absolutely no difference in the learning ability of the three groups. He then ran a very complicated learning set problem, depriving up to 47 hours, but even that group failed to be inferior to the other groups at the .06 confidence level. Had he tested human babies instead of monkeys, he would have found the same results, but this would not have been science and certainly would not have been socially acceptable. The experimenter was the only man in our laboratory who did not believe this data.

When the Hullian group wished to demonstrate the importance of primary drives, particularly hunger, they engaged in one of the limitless acts of legerdemain. They abandoned all of the time honored rodent learning techniques, particularly those involving any complex learning. Thus, the 14 blind alley maze became a matter of history only and probably only a subject for archeology. The maze they adopted became simpler and simpler and eventually became only a single straight alley. In this apparatus, for all practical purposes, the rat got nowhere and neither did the experimenter. Learning a problem became almost devoid of learning and, since practically no learning was involved, the speed of peregrination was influenced positively by deprivation. This was a very effective apparatus, if the desire was to prove the importance of the internal, primary drive. The apparatus tested basically the deprivation effect on nonlearning. It registered the speed of response and how rapidly a Ph.D. might be achieved.

Shortly before this study, we deliberately fed our monkeys food before we tested them, and, even though this made some theorists unhappy, it made the monkeys very happy, indeed. Their performance improved. An effective fact is better than two ineffective theories.

The world outside has multiple ways of stimulating the organism and these stimulating agents are described as incentives. Incentives started out in simple and dastardly dull fashion as tasty tidbits or a fragrant, fruit-flavored libation. Thanks in large part to monkey motives, new and drastically different incentives appeared, and although most psychologists resisted the influx of incentives for years, the decadent and destitute psychological scientists eventually relented. Fortunately, American motivational theory escaped from complete visceral vacuity.

Incentives not only increased the understanding of the many, many complex learning situations but introduced the principle of the variation of the strength of certain incentives in different combinations of incentives.

Faced with high incentive foods, the rhesus showed no social stimulation from the presence simultaneously of other monkeys. Monkeys eat as quickly as possible when the foods are flowery and fragrant and another monkey adds no increment to the motivation. Contrariwise, social facilitation may be enormous when two monkeys face low incentive food (Harlow & Yudin, 1933). The lonely monkey has no interest, but the paired monkeys' ingestion is greatly enhanced. This illustrates the fundamental fact that most primate fundamental motives are evil.

When the majority of psychologists were hypnotized by the primary motives of hunger, thirst, fear, sex, and pain, they tended to ignore complex behaviors, such as aggression and play, or else tuck these complex behaviors away in the roomy but quite undefinitive cupboard called learning. They also lost their curiosity, and this made it difficult for them to conceive of the motive of curiosity or exploration with all the different incentives involved.

For eons and eons it had been known that monkeys and apes and even human children would pursue and play with mechanical toys and puzzles. As a matter of fact, they would play with practically anything that was new and different, as long as the environment was familiar; and even if the environment was unfamiliar they would play with anything as long as it was muddy or gooey enough. In the main, these precise and persistent behaviors did not require the presence of food incentives nor dire and dank deprivation conditions. Obviously, incentives can be more than food or fluids. More often than not, such minor variables as food or fluids by their presence merely obscured meaningful motivation. Curiosity is one of the many maturational mechanisms linking monkeys to man.

Conventional incentive psychologists were not puzzled by puzzles nor piqued by people. They were merely uninterested. One of the father figures of modern psychological theory, Robert Woodworth, in his very last book, before the middle of the 20th century, gave full importance to the incentives, including those of curiosity and exploration. Perhaps because of his advanced years his internal motives had become suborned and he turned his attention to external incentives. Had lack of curiosity beset Archimedes and others of his ilk, the world would be a very different place today. Maybe it is. And, as Archimedes so succinctly put it, "Give me a lever and I can move the world, everything, that is, excepting the lever itself."

Sharing the importance of the lever, Archimedes and Skinner together could probably have moved the San Andreas fault. That might have prevented some psychologists from falling into it.

Faced with mechanical puzzles and toys, primates persist in attempted solutions more persistently and more efficiently if no physical incentives such as food are offered. Food apparently can be a distraction to thought. The animal becomes so eager for the food that it forgets its place in the solution of the puzzle.

The most dramatic and persistent example of a curiosity motive in rhesus monkeys was described by Butler (1953). He was trying, by the use of an obliquely positioned glass shield, to directly observe monkeys while they waited out the delay period in a delayed response test. Suddenly he realized, to his surprise, that the monkeys were spending more time observing him than he was spending observing them. Instantly, he realized that the monkeys either had an enormously powerful curiosity motive, or else they had none. He drew the right conclusion and created a monkey visual curiosity box.

In the curiosity box, the monkey could press open one of two doors built in the monkey's living cage and look out for 10 seconds before the door

Fig. 26. The Butler curiosity box.

automatically closed. Even this interruption did not inhibit monkey curiosity (see Fig. 26).

To test the persistence of monkey visual curiosity, Butler started one afternoon to see how long a monkey would seek to see. According to his objective data, a prize monkey continued to exhibit visual curiosity for 18 consecutive hours. Analysis of the data suggests that the monkeys continued to exhibit visual curiosity, but that the experimenter had fallen asleep.

A group of eight monkeys was tested on a simple 3-device mechanical puzzle apparatus presented unassembled but assembled and reassembled by the experimenters for half of the group, called the "A" subjects. For the other half, the "B" subjects, the puzzle was continuously left unassembled. This gave the B subjects a chance to assemble the puzzle themselves. Of course none of them succeeded. None of them tried. Indeed, the testers themselves could reassemble the devices only after specific description of assemblage instructions.

Any sensory system functioning correctly leads to manipulation. If you look at the primate, and sometimes I don't want to, take a look at his sensory system. Sensory systems are primarily vision. Most primates are visual animals, unless they're loris or lemurs. If they're loris or lemurs, there's no point in worrying about it. The visual system is an enormously complex system, as judged by both the central and the peripheral mechanisms. Visual exploration is a form of curiosity, and manipulation is the repsonse. There is no question but what curiosity is related to the incentive.

On days 13 and 14 of the experiment, the monkeys were observed during 5-minute periods, 1 hour apart, and the efficiency of their manipulation learning measured. All of the monkeys had learned how to disassemble the puzzle and all but one had improved rapidly. This one monkey would not solve the problem in the experimenter's presence but did so immediately after the experimenter moved on to the next cage. Apparently he was opposed to training experimenters. Throughout the experiment, when the monkeys completely unassembled the puzzle, they all stopped. There is obviously a motivating factor over and above the exploration and manipulation. The completion of the task, the accomplishment itself, is a reinforcement. The group B monkeys, who had not had the disassembling practice, made many errors and very few correct responses on the same 2 days of testing but proved to be willing subjects. During this entire experiment no food incentives were introduced at any time. Subsequently, however, the effect of food was explored.

In the presence of the monkey, food was baited underneath the hasp of the 3-device puzzle, and its effect on the previously learned manipulative motives was appraised. All the food incentive did was to enormously disrupt the behaviors learned through the power of the manipulation incentives by themselves. The number of errors vastly increased and no monkey ever learned to solve the problem as appropriately as before, through the most efficient manipulation of the devices. In the test without food reward, the monkeys always followed the most efficient sequences and practically never touched the hasp first. After being shown the food, the rhesus became hasp fixated or fastened or fascinated. Random movements and simple learned behaviors, which had no connection with solution of the puzzle and acquisition of the food were predominant. It has long been known that simple behaviors such as running down a straight alley are facilitated by food incentives and hunger and thirst deprivation states, but there is every indication that the opposite holds true for more complex learned behaviors.

Another very simple demonstration of curiosity and manipulation learning was provided by the use of a wooden panel in which Harlow placed 10 screw eyes in 2 vertical rows. The red screw eyes were easily removed; the green screw eyes were immobile and the monkey was left to his own resources. Very rapidly the monkeys learned to pull out the obliging red screw eyes and to totally ignore the recalcitrant green ones. This is far from a complex learning situation, but does illustrate achievement through discrimination, motivated by both the manipulation and the fact of accomplishment. It also makes understandable the number of teddy bears floating around the world minus their button eyes.

The effect of brain damage on visual curiosity in monkeys was also measured by the frequency and persistence of responses after recovery from various types of brain damage. The results indicate that all of the animals, both brain-damaged and normal, responded effectively in the Butler box and that they were persistent over a period of hours. The curiosity box manipulation was successfully mastered by most of the animals but there was no evidence of progressive learning among the operated animals, a tribute to the strength of the manipulation motive and to the requirements of learning.

A dramatic and disastrous illustration of the power of curiosity in an anthropoid ape is the hopeless story of Jiggs, the nicest Orangutan that ever lived in a cage.

To alleviate Jiggs' cage confines and long-term loneliness, we gave Jiggs some playthings: a wooden block with a square hole and a square plunger, and another wooden block with a round hole and a round plunger. Jiggs attacked the potential problem with insatiable curiosity and, after a time, he learned to insert the round plunger into both the round hole and the square hole. Eventually, he learned to insert the square plunger into the square hole, but, to his eternal disgust, he could not insert the square plunger into the round hole.

Had he succeeded, this would have been his greatest triumph; but the true triumph of this period was the social rehabilitation of Jiggs' wife, Maggie. Maggie and Jiggs received their nomenclature because day in and day out Maggie beat Jiggs unmercifully, probably just for practice. Like his comic strip namesake, Jiggs had, in some way or other, learned that it was better to receive than to give.

When Maggie was a little younger, Mr. Winkleman, the Director of the Madison Zoo, used to take her out and show her the viable virtues of Vilas Park. For months, everything went according to plan, but then, all at once, during an outing, a little urchin, without urgin', picked up a handy rock and threw it hard at Maggie's forehead. Maggie drew back with every intent of committing meaningful mayhem. Winkleman was horrified because he was convinced that Maggie would dismember or kill the urchin and there are never really enough to go around. He looked around and there at his feet was a real baseball bat, which he picked up and with a swatter's swing struck Maggie full force across the forehead. Maggie never fell, she never cried out. She merely rubbed the forehead with the back of her hand, which she then extended to Mr. Winkleman and followed him gently back to her cage. Apparently, after all these years, she had finally found the man of her dreams, a man who understood the psychology of women.

CHAPTER 6

MONKEYS, MICE, MEN, AND MOTIVES

Many of psychology's theoretical growing pains—or, in modern terminology, conditioned anxieties—stem from the behavioral revolution of Watson. The new psychology intuitively disposed of instincts and painlessly disposed of hedonism. But having completed this St. Bartholomew-type massacre, behavioristic motivation theory was left with an aching void, a nonhedonistic aching void, needless to say.

Before the advent of the Watsonian scourge, the importance of external stimuli as motivating forces was well recognized. Psychologists will always remain indebted to Loeb's (1918) brilliant formulation of tropistic theory, which emphasized, and probably overemphasized, the powerful role of external stimulation as the primary motivating agency in animal behavior. Unfortunately, Loeb's premature efforts to reduce all behavior to overly simple mathematical formulation, his continuous acceptance of new tropistic constructs in an effort to account for any aberrant behavior not easily integrated into his original system, and his abortive attempt to encompass all behavior into a miniature theoretical system doubtless led many investigators to underestimate the value of his experimental contributions.

Thorndike (1911) was simultaneously giving proper emphasis to the role of external stimulation as a motivating force in learning and learned performances. Regrettably, these motivating processes were defined in terms of pain and pleasure, and it is probably best for us to dispense with such lax, ill-defined, subjective terms as pain, pleasure, anxiety, frustration, and hypotheses—particularly in descriptive and theoretical rodentology.

Instinct theory, for all its terminological limitations, put proper emphasis on the motivating power of external stimuli; for, as so brilliantly described by Watson (1914), the instinctive response was elicited by "serial stimulation," much of which was serial external stimulation.

The almost countless researches on tropisms and instincts might well have been expanded to form a solid and adequate motivational theory for psychology—a theory with a proper emphasis on the role of the external stimulus and an emphasis on the importance of incentives as opposed to internal drives per se.

It is somewhat difficult to understand how this vast and valuable literature was to become so completely obscured and how the importance of the external stimulus as a motivating agent was to become lost. Pain-pleasure theory was discarded because the terminology had subjective, philosophical implications. Instinct theory fell into disfavor because psychologists rejected the dichotomized heredity-environment controversy and, also, because the term "instinct" had more than one meaning. Why tropistic theory disappeared remains a mystery, particularly inasmuch as most of the researches were carried out on subprimate animal forms.

Modern motivation theory apparently evolved from an overpopularization of certain experimental and theoretical materials. Jennings' (1906) demonstration that "physiological state" played a role in determining the behavior of the lower animal was given exaggerated importance and emphasis, thereby relegating the role of external stimulation to a secondary position as a force in motivation. The outstanding work in the area of motivation between 1920 and 1930 related to visceral drives and drive cycles, and was popularized by Richter's idealized theoretical paper on "Animal Behavior and Internal Drives" (1927) and Cannon's *The Wisdom of the Body* (1932).

When the self-conscious behavior theorists of the early thirties looked for a motivation theory to integrate with their developing learning constructs, it was only natural that they should choose the available tissue-tension hypothesis. Enthusiastically and uncritically, the S-R theorists swallowed these theses whole. For 15 years they have tried to digest them, and it is now time that these theses be subjected to critical examination, analysis, and evaluation. We do not question that these theses have fertilized the field of learning, but we do question that the plants that have developed are those that will survive the test of time.

It is my belief that the theory which describes learning as dependent upon drive reduction is false, that internal drive as such is a variable of little importance to learning, and that this small importance steadily decreases as we ascend the phyletic scale and as we investigate learning problems of progressive complexity. Finally, it is my position that drive-reduction theory orients learning psychologists to attack problems of limited importance and to ignore the fields of research that might lead us, in some foreseeable future time, to evolve a theoretical psychology of learning that transcends any single species or order.

There can be no doubt that the single-celled organisms, such as the amoeba and the paramecium, are motivated to action both by external and internal stimuli. The motivation by external stimulation gives rise to heliotropisms, chemotropisms, and rheotropisms. The motivation by internal stimulation produces characteristic physiological states which have, in turn, been described as chemotropisms. From a phylogenetic point of view, moreover, neither type of motive appears to be more basic or more fundamental than the other. Both types are found in the simplest known animals and function in interactive, rather than in dominant-subordinate, roles.

Studies of fetal responses in animals, from opossum to man, give no evidence suggesting that the motivation of physiological states precedes that of external incentives. Tactual, thermal, and even auditory and visual stimuli elicit complex patterns of behavior in the fetal guinea pig, although this animal has a placental circulation which should guarantee against thirst or hunger (Carmichael, 1934). The newborn opossum climbs up the belly of the female and into the pouch, apparently in response to external cues; if visceral motives play any essential role, it is yet to be described (Langworthy, 1928). The human fetus responds to external tactual and nociceptive stimuli at a developmental period preceding demonstrated hunger or thirst motivation. Certainly, there is no experimental literature to indicate that internal drives are ontogenetically more basic than exteroceptive motivating agencies.

Tactual stimulation, particularly of the cheeks and lips, elicits mouth, head, and neck responses in the human neonate, and there are no data demonstrating that these responses are conditioned, or even dependent, upon physiological drive states. Hunger appears to lower the threshold for these tactual stimuli. Indeed, the main role of the primary drive seems to be one of altering the threshold for precurrent responses. Differentiated sucking response patterns have been demonstrated to quantitatively varied thermal chemical stimuli in the infant only hours of age (Jensen, 1932), and there is, again, no reason to believe that the differentiation could have resulted from antecedent tissue-tension reduction states. Taste and temperature sensations induced by the temperature and chemical composition of the liquids seem adequate to account for the responses.

There is neither phylogenetic nor ontogenetic evidence that drive states elicit more fundamental and basic response patterns than do external stimuli; nor is there basic for the belief that precurrent responses are more dependent upon consummatory responses than are consummatory responses dependent upon precurrent responses. There is no evidence that the differentiation of the inate precurrent responses is more greatly influenced by tissue-tension reduction than are the temporal ordering and intensity of consummatory responses influenced by conditions of external stimulation.

There are logical reasons why a drive-reduction theory of learning, a theory which emphasizes the role of internal, physiological-state motivation, is entirely untenable as a motivational theory of learning. The internal drives are cyclical and operate, certainly at any effective level of intensity, for only a brief fraction of any organism's waking life. The classical hunger drive physiologically defined ceases almost as soon as food—or nonfood—is ingested. This, as far as we know, is the only case in which a single swallow portends anything of importance. The temporal brevity of operation of the internal drive states obviously offers a minimal opportunity for conditioning and a maximal opportunity for extinction. The human being, at least in the continental United States, may go for days or even years without ever experiencing true hunger or thirst. If his complex conditioned responses were dependent upon primary drive reduction, one would expect him to regress rapidly to a state of tuitional oblivion. There are, of course, certain recurrent physiological drive states that are maintained in the adult. But the studies of Kinsey, Pomeroy, and Martin (1948) indicate that in the case of one of these there is an inverse correlation between presumed drive strength and scope and breadth of learning; and in spite of the alleged reading habits of the American public, it is hard to believe that the other is our major source of intellectual support. Any assumption that derived drives or motives can account for learning in the absence of primary drive reduction puts an undue emphasis on the strength and permanence of derived drives, at least in subhuman animals. Experimental studies to date indicate that most derived drives (Miller, 1951) and second-order conditioned responses (Pavlov, 1927) rapidly extinguish when the rewards which theoretically reduce the primary drives are withheld. The additional hypothesis of functional autonomy of motives, which could bridge the gap, is yet to be demonstrated experimentally.

The condition of strong drive is inimical to all but very limited aspects of learning—the learning of ways to reduce the internal tension. The hungry child screams, closes his eyes, and is apparently oblivious to most of his environment. During this state he eliminates response to those aspects of his environment around which all his important learned behaviors will be based. The hungry child is a most incurious child, but after he has eaten and become thoroughly sated, his curiosity and all the learned responses associated with his curiosity take place. If this learning is conditioned to an internal drive state, we must assume it is the resultant of backward conditioning. If we wish to hypothesize that backward conditioning is dominant over forward conditioning in the infant, it might be possible to reconcile fact with S-R theory. It would appear, however, that alternate theoretical possibilities should be explored before the infantile backward conditioning hypothesis is accepted.

Observations and experiments on monkeys convinced us that there was as much evidence to indicate that a strong drive state inhibits learning as to indicate that it facilitates learning. It was the speaker's feeling that monkeys learned most efficiently if they were given food before testing, and, as a result, the speaker routinely fed his subjects before every training session. The rhesus monkey is equipped with enormous cheek pouches, and consequently many subjects would begin the educational process with a rich store of incentives crammed into the buccal cavity. When the monkey made a correct response, it would add a raisin to the buccal storehouse and swallow a little previously munched food. Following an incorrect response, the monkey would also swallow a little stored food. Thus, both correct and incorrect responses invariably resulted in S-R theory drive reduction. It is obvious that under these conditions the monkey cannot learn, but I developed an understandable skepticism of this hypothesis when the monkeys stubbornly persisted in learning, learning rapidly, and learning problems of great complexity. Because food was continuously available in the monkey's mouth, an explanation in terms of differential fractional anticipatory goal responses did not appear attractive. It would seem that the Lord was simply unaware of drive-reduction learning theory when he created, or permitted the gradual evolution of, the rhesus monkey.

The langurs are monkeys that belong to the only family of primates with sacculated stomachs. There would appear to be no mechanism better designed than the sacculated stomach to induce automatically prolonged delay of reinforcement defined in terms of homeostatic drive reduction. Langurs should, therefore, learn with great difficulty. But a team of Wisconsin students has discovered that the langurs in the San Diego Zoo learn at a high level of monkey efficiency. There is, of course, the alternative explanation that the inhibition of hunger contractions in multiple stomachs is more reinforcing than the inhibition of hunger contractions in one. Perhaps the quantification of the gastric variable will open up great new vistas of research.

Actually, the anatomical variable of diversity of alimentary mechanisms is essentially uncorrelated with learning to food incentives by monkeys and suggests that learning efficiency is far better related to tensions in the brain than in the belly.

Experimental test bears out the fact that learning performance by the monkey is unrelated to the theoretical intensity of the hunger drive. Meyer (1951) tested rhesus monkeys on discrimination-learning problems under conditions of maintenance-food deprivation of 1.5, 18.5, and 22.5 hours and found no significant differences in learning or performance. Subsequently, he tested the same monkeys on discrimination-reversal learning following 1, 23, and 47 hours of maintenance-food deprivation and, again, found no significant differences in learning or in performance as measured by activity, direction of activity, or rate of responding. There was some evidence, not statistically significant, that the most famished subjects were a bit overeager and that intense drive exerted a mildly inhibitory effect on learning efficiency.

Meyer's data are in complete accord with those presented by Birch (1945) who tested six young chimpanzees after 2, 6, 12, 24, and 48 hours of food deprivation and found no significant differences in proficiency of performance on six patterned string problems. Observational evidence led Birch to conclude that intense food deprivation adversely affected problem solution, because it led the chimpanzee to concentrate on the goal to the relative exclusion of the other factors.

It may be stated unequivocally that, regardless of any relationship that may be found for other animals, there are no data indicating that intensity of drive state and the presumably correlated amount of drive reduction are positively related to learning efficiency in primates.

In point of fact, there is no reason to believe that the rodentological data will prove to differ significantly from those of monkey, chimpanzee, and man. Strassburger (1950) has recently demonstrated that differences in food deprivation from 5 hours to 47 hours do not differentially affect the habit strength of the bar-pressing response as measured by subsequent resistance to extinction. Recently, Sheffield and Roby (1950) have demonstrated learning in rats in the absence of primary drive reduction. Hungry rats learned to choose a maze path leading to a saccharin solution, a nonnutritive substance, in preference to a path leading to water. No study could better illustrate the predominant role of the external incentive-type stimulus on the learning function. These data suggest that, following the example of the monkey, even the rats are abandoning the sinking ship of reinforcement theory.

The effect of intensity of drive state on learning doubtless varies as we ascend the phyletic scale and certainly varies, probably to the point of almost complete reversal, as we pass from simple to complex problems, a point emphasized some years ago in a theoretical article by Maslow (1943). Intensity of painful stimulation prevents the monkey from solving any problem of moderate complexity. This fact is consistent with a principle that was formulated and demonstrated experimentally many years ago as the Yerkes-Dodson law (1908). There is, of course, no reference to the Yerkes-Dodson law by any drive-reduction theorist.

We do not mean to imply that drive state and drive-state reduction are unrelated to learning; we wish merely to emphasize that they are relatively unimportant variables. Our primary quarrel with drive-reduction theory is that it tends to focus more and more attention on problems of less and less importance. A strong case can be made for the proposition that the importance of the psychological problems studied during the last 15 years has decreased as a negatively accelerated function approaching an asymptote of complete indifference. Nothing better illustrates this point than the kinds of apparatus currently used in "learning" research. We have the single-unit T-maze, the straight runway, the double-compartment grill box, and the Skinner box. The single-unit T-maze is an ideal apparatus for studying the visual capacities of a nocturnal animal; the straight runway enables one to measure quantitatively the speed and rate of running from one dead end to another; the double-compartment grill box is without doubt the most efficient torture chamber which is still legal; and the Skinner box enables one to demonstrate discrimination learning in a greater number of trials than is required by any other method. But the apparatus, though inefficient, give rise to data which can be splendidly quantified. The kinds of learning problems which can be efficiently measured in these apparatus represent a challenge only to the decorticate animal. It is a constant source of bewilderment to me that the neobehaviorists who so frequently belittle physiological psychology should choose apparatus which, in effect, experimentally decorticate their subjects.

The Skinner box is a splendid apparatus for demonstrating that the rate of performance of a learned response is positively related to the period of food deprivation. We have confirmed this for the monkey by studying rate of response on a modified Skinner box following 1, 23, and 47 hours of food deprivation. Increasing length of food deprivation is clearly and positively related to increased rate of response. This functional relationship between drive states and responses does not hold, as we have already seen, for the monkey's behavior in discrimination learning or in acquisition of any more complex problem. The data, however, like rate data, are in complete accord with Crozier's (1929) finding that the acuteness of the radial angle of tropistic movement in the slug Limax is positively related to intensity of the photic stimulation. We believe there is generalization in this finding, and we believe the generalization to be that the results from the investigation of simple behavior may be very informative about even simpler behavior but very seldom are they informative about behavior of greater complexity. I do not want to discourage anyone from the pursuit of the behavior of the slug or even the psychological Holy Grail by the use of the Skinner box, but as far as I am concerned, there will be no moaning of farewell when we have passed the pressing of the bar.

In the course of human events many psychologists have children, and these children always behave in accord with the theoretical position of their parents. For purposes of scientific objectivity, the boys are always referred to as "Johnny" and the girls as "Mary." For some 11 months, I have been observing the behavior of Mary X. Perhaps the most striking characteristic of this particular primate has been the power and persistence of her curiosity-investigatory motives. At an early age Mary X demonstrated a positive valence to parental thygmotatic stimulation. My original interpretation of these tactual-thermal erotic responses as indicating parental affection was dissolved by the discovery that when Mary X was held in any position depriving her of visual exploration of the environment, she screamed; when held in a position favorable to visual exploration of the important environment, which did not include the parent, she responded positively. With the parent and position held constant and visual exploration denied by snapping off the electric light, the positive responses changed to negative, and they returned to positive when the light was again restored. This behavior was observed in Mary X, who, like any good Watson child, showed no "innate fear of the dark."

The frustrations of Mary X appeared to be, in large part, the results of physical inability to achieve curiosity-investigatory goals. In her second month, frustrations resulted from inability to hold up her head indefinitely while lying prone in her crib or on a mat and the consequent loss of visual curiosity goals. Each time she had to lower her head to rest, she cried lustily. At 9 weeks, attempts to explore (and destroy) objects anterior resulted in wriggling backward away from the lure and elicited violent negative responses. Once she negotiated forward locomotion, exploration set in, in earnest, and, much to her parents' frustration, showed no sign of diminishing.

Can anyone seriously believe that the insatiable curiosity-investigatory motivation of the child is a second-order or derived drive conditioned upon hunger or sex or any other internal drive? The S-R theorist and the Freudian psychoanalyst imply that such behaviors are based on primary drives. An informal survey of neobehaviorists who are also fathers (or mothers) reveals that all have observed the intensity and omnipresence of the curiosity-investigatory motive in their own children. None of them seriously believes that the behavior derives from a second-order drive. After describing their children's behavior, often with a surprising enthusiasm and frequently with the support of photographic records, they trudge off to their laboratories to study, under conditions of solitary confinement, the intellectual processes of rodents. Such attitudes, perfectly in keeping with drive-reduction theory, no doubt account for the fact that there are no experimental or even systematic observational studies of curiosity-investigatory-type external-incentive motives in children.

A key to the real learning theory of any animal species is knowledge of the nature and organization of the unlearned patterns of response. The differences in the intellectual capabilities of cockroach, rat, monkey, chimpanzee, and man are as much a function of the differences in the inherent patterns of response and the differences in the inherent motivational forces as they are a function of sheer learning power. The differences in these inherent patterns of response and in the motivational forces will, I am certain, prove to be differential responsiveness to external stimulus patterns. Furthermore, I am certain that the variables which are of true, as opposed to psychophilosophical, importance are not constant from learning problem to learning problem even for the same animal order, and they are vastly deverse as we pass from one animal order to another.

Convinced that the key to human learning is not the conditioned response but, rather, motivation aroused by external stimuli, the speaker has initiated researches on curiosity-manipulation behavior as related to learning in monkeys (Davis, Settlage, & Harlow, 1950; Harlow, 1950; Harlow, Harlow, & Meyer, 1950). The justification for the use of monkeys is that we have more monkeys than children. Furthermore, the field is so unexplored that a systematic investigation anywhere in the phyletic scale should prove of methodological value. The rhesus monkey is actually a very incurious and nonmanipulative animal compared with the anthropoid apes, which are, in turn, very incurious nonmanipulative animals compared with man. It is certainly more than coincidence that the strength and range of curiosity-manipulative motivation and position within the primate order are closely related.

We have presented three studies which demonstrate that monkeys can and do learn to solve mechanical puzzles when no motivation is provided, other than presence of the puzzle. Furthermore, we have presented data to show that, once mastered, the sequence of manipulation involved in solving these puzzles is carried out relatively flawlessly and extremely persistently. We have presented what we believe is incontrovertible evidence against a second-order drive interpretation of this learning.

A fourth study was carried out recently by Gately at the Wisconsin laboratories. Gately directly compared the behavior of two groups of four monkeys presented with banks of four identical mechanical puzzles, each utilizing three restraining devices. All four food-plus-puzzle-rewarded monkeys solved the four identical puzzles, and only one of the four monkeys, motivated by curiosity alone, solved all the puzzles. This one monkey, however, learned as rapidly and as efficiently as any of the food-rewarded monkeys. But I wish to stress an extremely important observation made by Gately and supported by quantitative records. When the food-rewarded monkeys had solved a puzzle, they abandoned it. When the nonfood-rewarded animals had solved the puzzle, they frequently continued their explorations and manipulations. Indeed, one reason for the nonfood-rewarded monkeys' failure to achieve the experimenter's concept of solution lay in the fact that the monkey became fixated in exploration and manipulation of limited puzzle or puzzle-device components. From this point of view, hunger-reduction incentives may be regarded as motivation-destroying, not motivation-supporting, agents.

Twenty years ago at the Vilas Park Zoo, in Madison, we observed an adult orangutan given two blocks of wood, one with a round hole, one with a square hole, and two plungers, one round and one square. Intellectual curiosity alone led it to work on these tasks, often for many minutes at a time, and to solve the problem of inserting the round plunger in both holes. The orangutan never solved the problem of inserting the square peg into the round hole, but inasmuch as it passed away with perforated ulcers a month after the problem was presented, we can honestly say that it died trying. And in defense of this orangutan, let it be stated that it died working on more complex problems than are investigated by most present-day learning theorists.

Schiller has reported that chimpanzees solve multiple-box-stacking problems without benefit of food rewards, and he had presented observational evidence that the joining of sticks resulted from manipulative play responses.

The cebus monkey has only one claim to intellectual fame—an ability to solve instrumental problems that rivals the much publicized ability of the anthropoid apes (Harlow, 1951; Klüver, 1933). It can be no accident that the cebus monkey, inferior to the rhesus on conventional learning tasks, demonstrates far more spontaneous instrumental-manipulative responses than any old-world form. The complex, innate external-stimulus motives are variables doubtlessly as important as, or more important than, tissue tensions, stimulus generalization, excitatory potential, or secondary reinforcement. It is the oscillation of sticks, not cortical neurons, that enables the cebus monkey to solve instrumental problems.

No matter how important may be the analysis of the curiosity-manipulative drives and the learning which is associated with them, we recognize the vast and infinite technical difficulties that are inherent in the attack on the solution of these problems—indeed, it may be many years before we can routinely order such experiments in terms of latin squares and factorial designs, the apparent sine qua non for publication in the *Journal of Experimental Psychology* and the *Journal of Comparative and Physiological Psychology*.

There is, however, another vast and important area of external-stimulus incentives important to learning which has been explored only superficially and which can, and should, be immediately and systematically attacked by rodentologists and primatologists alike. This is the area of food incentives—or, more broadly, visuo-chemo variables—approached from the point of view of their function as motivating agents per se. This function, as the speaker sees it, is primarily an affective one and only secondarily one of tissue-tension reduction. To dispel any fear of subjectivity, let us state that the affective tone of food incentives can probably be scaled by preference tests with an accuracy far exceeding any scaling of tissue tensions. Our illusion of the equal-step intervals of tissue tensions is the myth that length of the period of deprivation is precisely related to tissue-tension intensity, but the recent experiments by Koch and Daniel (1945) and Horenstein (1951) indicate that this is not true, thus beautifully confirming the physiological findings of 30 years ago.

Paired-comparison techniques with monkeys show beyond question that the primary incentive variables of both differential quantity and differential quality can be arranged on equal-step scales, and there is certainly no reason to believe that variation dependent upon subjects, time, or experience is greater than that dependent upon physiological hunger.

In defense of the rat and its protagonists, let it be stated that there are already many experiments on this lowly mammal which indicate that its curiosity-investigatory motives and responsiveness to incentive variables can be quantitatively measured and their significant relationship to learning demonstrated. The latent learning experiments of Buxton (1940), Haney (1931), Seward, Levy, and Handlon (1950) and others have successfully utilized the exploratory drive of the rat. Keller (1941) and Zeaman and House (1950) have utilized the rat's inherent aversion to light, or negative heliotropistic tendencies, to induce learning. Flynn and Jerome (1952) have shown that the rat's avoidance of light is an external-incentive motivation that may be utilized to obtain the solution of complex learned performances. For many rats it is a strong and very persistent form of motivation. The importance of incentive variables in rats has been emphasized and reemphasized by Young (1949), and the influence of incentive variables on rat learning has been demonstrated also by Zeaman (1949), Crespi (1942), and others. I am not for one moment disparaging the value of the rat as a subject for psychological investigation; there is very little wrong with the rat that cannot be overcome by the education of the experimenters.

It may be argued that if we accept the theses of this chapter, we shall be returning to an outmoded psychology of tropisms, instincts, and hedonism. There is a great deal of truth to this charge. Such an approach might be a regression were it not for the fact that psychology now has adequate techniques of methodology and analysis to attack quantifiably these important and neglected areas. If we are ever to have a comprehensive theoretical psychology, we must attack the problems whose solution offers hope of insight into human behavior, and it is my belief that if we face our problems honestly and without regard to, or fear of, difficulty, the theoretical psychology of the future will catch up with, and eventually even surpass, common sense.

CHAPTER 7

THE NATURE OF LOVE

Love is a wondrous state, deep, tender, and rewarding. Because of its intimate and personal nature it is regarded by some as an improper topic for experimental research. But, whatever our personal feelings may be, our assigned mission as psychologists is to analyze all facets of human and animal behavior into their component variables. So far as love or affection is concerned, psychologists have failed in this mission. The little we know about love does not transcend simple observation, and the little we write about it has been written better by poets and novelists. But of greater concern is the fact that psychologists tend to give progressively less attention to a motive which pervades our entire lives. Psychologists, at least psychologists who write textbooks, not only show no interest in the origin and development of love or affection, but they seem to be unaware of its very existence.

The apparent repression of love by modern psychologists stands in sharp contrast with the attitude taken by many famous and normal people. The word "love" has the highest reference frequency of any word cited in Bartlett's book *Familiar Quotations*. It would appear that this emotion has long had a vast interest and fascination for human beings, regardless of the attitude taken by psychologists; but the quotations cited, even by famous and normal people, have a mundane redundancy. These authors and authorities have stolen love from the child and infant and made it the exclusive property of the adolescent and adult.

Thoughtful men, and probably all women, have speculated on the nature of love. From the developmental point of view, the general plan is quite clear: The initial love responses of the human being are those made by the infant to the mother or some mother surrogate. From this intimate attachment of the child to the mother, multiple learned and generalized affectional responses are formed.

Unfortunately, beyond these simple facts we know little about the fundamental variables underlying the formation of affectional responses and little about the mechanisms through which the love of the infant for the mother develops into the multifaceted response patterns characterizing love or affection in the adult. Because of the dearth of experimentation, theories about the fundamental nature of affection have evolved at the level of observation, intuition, and discerning guesswork, whether these have been proposed by psychologists, sociologists, anthropologists, physicians, or psychoanalysts.

The position commonly held by psychologists and sociologists is quite clear: The basic motives are, for the most part, the primary drives—particularly hunger, thirst, elimination, pain, and sex—and all other motives, including love or affection, are derived or secondary drives. The mother is associated with the reduction of the primary drives—particularly hunger, thirst, and pain—and through learning, affection or love is derived.

It is entirely reasonable to believe that the mother, through association with food, may become a secondary-reinforcing agent, but this is an inadequate mechanism to account for the persistence of the infant-maternal ties. There is a spate of researches on the formation of secondary reinforcers to hunger and thirst reduction. There can be no question that almost any external stimulus can become a secondary reinforcer if properly associated with tissue-need reduction, but the fact remains that this redundant literature demonstrates unequivocally that such derived drives suffer relatively rapid experimental extinction. Contrariwise, human affection does not extinguish when the mother ceases to have intimate association with the drives in question. Instead, the affectional ties to the mother show a lifelong, unrelenting persistence and, even more surprising, widely expanding generality.

Oddly enough, one of the few psychologists who took a position counter to modern psychological dogma was John B. Watson, who believed that love was an innate emotion elicited by cutaneous stimulation of the erogenous zones. But experimental psychologists, with their peculiar propensity to discover facts that are not true, brushed this theory aside by demonstrating that the human neonate had no differentiable emotions, and they confirmed, for the psychological profession, the fundamental law that prophets are without honor in their own country.

The psychoanalysts have concerned themselves with the problem of the nature of the development of love in the neonate and infant, using ill and aging human beings as subjects. They have discovered the overwhelming importance of the breast and related this to the oral erotic tendencies developed at an age preceding their subjects' memories. Their theories range from a belief that the infant has an innate need to achieve and suckle at the breast to beliefs not unlike commonly accepted psychological theories. There are exceptions, as seen in the recent writings of John Bowlby (1969), who attributes importance not only to food and thirst satisfaction, but also to "primary object-clinging," a need for intimate physical contact, which is initially associated with the mother.

As far as I know, there exists no direct experimental analysis of the relative importance of the stimulus variables determining the affectional or love responses in the neonatal and infant primate. Unfortunately, the human neonate is a limited experimental subject for such researchers, because or his inadequate motor capabilities. By the time the human infant's motor responses can be precisely measured, the antecedent determining conditions cannot be defined, having been lost in a jumble and jungle of confounded variables.

Many of these difficulties can be resolved by the use of the neonatal and infant macaque monkey as the subject for the analysis of basic affectional variables. It is possible to make precise measurements in this primate beginning at 2 to 10 days of age, depending upon the maturational status of the individual animal at birth. The macaque infant differs from the human infant in that the monkey is more mature at birth and grows more rapidly; but the basic responses relating to affection, including nursing, contact, clinging, and even visual and auditory exploration, exhibit no fundamental differences in the two species. Even the development of perception, fear, frustration, and learning capability follows very similar sequences in rhesus monkeys and human children.

Three years' experimentation before we started our studies on affection gave us concentrated experience with the neonatal monkey. We had separated more than 60 of these animals from their mothers, 6 to 12 hours after birth, and suckled them on tiny bottles. The infant mortality was only a small fraction of what would have been obtained had we let the monkey mothers raise their infants. Our bottle-fed babies were healthier and heavier than monkey-mother-reared infants. We know that we are better initial monkey mothers than are real monkey mothers, thanks to synthetic diets, vitamins, iron extracts, penicillin, chloromycetin, and 5% glucose, in addition to constant, tender, loving care.

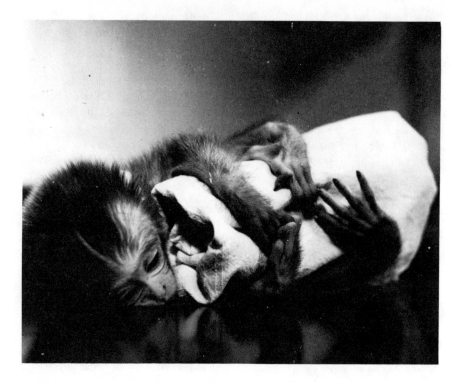

Fig. 27. Neonatal response to a cloth pad.

During the course of these studies we noticed that the laboratory-raised babies showed strong attachment to the cloth pads (folded gauze diapers) which were used to cover the hardware-cloth floors of their cages. The infants clung to these pads and engaged in violent temper tantrums when the pads were removed and replaced for sanitary reasons. Such contact-need or responsiveness had been reported previously by Gertrude van Wagenen (1950) for the monkey and by Thomas McCulloch and George Haslerud (1939) for the chimpanzee, and is reminiscent of the devotion often exhibited by human infants to their pillows, blankets, and soft, cuddly stuffed toys. Responsiveness by the 1-day-old infant monkey to the cloth pad is shown in Figure 27,

Fig. 28. Six-month old in gauze glory.

and an unusual and strong attachment of a 6-month-old infant to the cloth pad is illustrated in Figure 28. The baby, human or monkey, if it is to survive, must clutch at more than a straw.

We had also discovered, during some allied observational studies, that a baby monkey raised on a bare wire-mesh cage floor survives with difficulty, if at all, during the first 5 days of life. If a wire-mesh cone is introduced, the baby does better; and, if the cone is covered with terry cloth, husky, healthy, happy babies evolve. It takes more than a baby and a box to make a normal monkey. We were impressed by the possibility that, above and beyond the bubbling fountain of breast or bottle, contact comfort might be a very important variable in the development of the infant's affection for the mother.

At this point we decided to study the development of affectional responses of neonatal and infant monkeys to an artificial, inanimate mother, and so we built a surrogate mother which we hoped and believed would be a good surrogate mother. In devising this surrogate mother we were dependent neither upon the capriciousness of evolutionary processes nor upon mutations produced by chance radioactive fallout. Instead, we designed the mother surrogate in terms of modern human-engineering principles. We produced a perfectly proportioned, streamlined body stripped of unnecessary bulges and appendices. Redundancy in the surrogate mother's system was avoided by reducing the number of breasts from two to one and placing this unibreast in an upper-thoracic, sagittal position, thus maximizing the natural and known perceptual-motor capabilities of the infant operator. The surrogate was made from a block of wood, covered with sponge rubber, and sheathed in tan cotton terry cloth. A light bulb behind her radiated heat. The result was a mother, soft, warm, and tender, a mother with infinite patience, a mother available 24 hours a day, a mother that never scolded her infant and never struck or bit her baby in anger. Furthermore, we designed a mother-machine with maximal maintenance efficiency, since failure of any system or function could be resolved by the simple substitution of black boxes and new component parts. It is our opinion that we engineered a very superior monkey mother, although this position is not held universally by the monkey fathers.

Before beginning our initial experiment, we also designed and constructed a second mother surrogate, a surrogate in which we deliberately built less than the maximal capability for contact comfort. She is made of wire-mesh, a substance entirely adequate to provide postural support and nursing capability, and she is warmed by radiant heat. Her body differs in no essential way from that of the cloth mother surrogate, other than in the quality of the contact comfort which she can supply.

In our initial experiment, the dual mother-surrogate condition, a cloth mother and a wire mother were placed in different cubicles attached to the infant's living cage, as shown in Figure 29. For four newborn monkeys, the cloth mother lactated and the wire mother did not; and, for the other four, this condition was reversed. In either condition, the infant received all its milk through the mother surrogate as soon as it was able to maintain itself in this way, a capability achieved within 2 or 3 days, except in the case of very immature infants. Supplementary feedings were given until the milk intake from the mother surrogate was adequate. Thus, the experiment was designed as a test of the relative importance of the variables of contact comfort and nursing comfort. During the first 14 days of life, the monkey's cage floor was covered with a heating pad wrapped in a folded gauze diaper, and thereafter

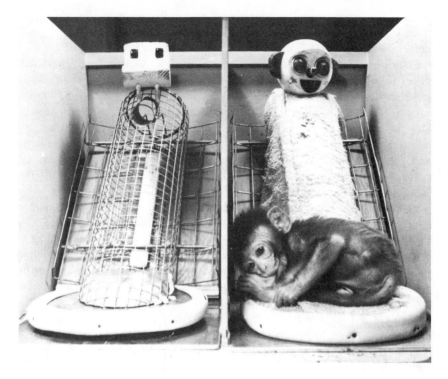

Fig. 29. Wire and cloth mother surrogates.

TIME SPENT ON DUAL MOTHERS

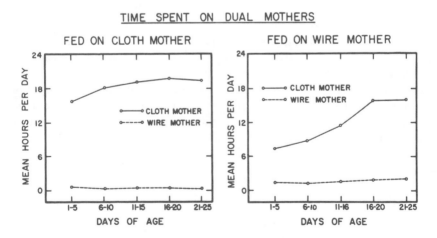

Fig. 30. Time spent on cloth and wire mother surrogates.

the cage floor was bare. The infants were always free to leave the heating pad or cage floor to contact either mother, and the time spent on the surrogate mothers was automatically recorded. Figure 30 shows the total time spent on the cloth and wire mothers under the two conditions of feeding. These data make it obvious that contact comfort is a variable of overwhelming importance in the development of affectional responses, whereas lactation is a variable of negligible importance. With age and opportunity to learn, subjects with the lactating wire mother showed decreasing responsiveness to her and increasing responsiveness to the nonlactating cloth mother, a finding completely contrary to any interpretation of derived drive in which the mother-form becomes conditioned to hunger-thirst reduction. These differential responses persisted throughout 165 consecutive days of testing.

One control group of neonatal monkeys was raised on a single wire mother, and a second control group was raised on a single cloth mother. There were no differences between these two groups in amount of milk ingested or in weight gain. The only difference between the groups lay in the composition of the feces, the softer stools of the wire-mother infants suggesting psychosomatic involvement. The wire mother is biologically adequate but psychologically inept.

We were not surprised to discover that contact comfort was an important basic affectional or love variable, but we did not expect it to overshadow so completely the variable of nursing; indeed, the disparity is so great as to suggest that the primary function of nursing as an affectional variable is that of insuring frequent and intimate body contact of the infant with the mother. Certainly, man cannot live by milk alone. Love is an emotion that does not need to be bottle- or spoon-fed, and we may be sure that there is nothing to be gained by giving lip service to love.

A charming lady once heard me describe these experiments; and, when I subsequently talked to her, her face brightened with sudden insight: "Now I know what's wrong with me," she said, "I'm just a wire mother." Perhaps she was lucky. She might have been a wire wife.

We believe that contact comfort has long served the animal kingdom as a motivating agent for affectional responses. Since at the present time we have no experimental data to substantiate this position, we supply information which must be accepted, if at all, on the basis of face validity:

THE SNAKE

To baby vipers, scaly skin
Engenders love 'twixt kith and kin.
Each animal by God is blessed
With kind of skin it loves the best.

THE HIPPOPOTAMUS

This is the skin some babies feel
Replete with hippo love appeal.
Each contact, cuddle, push, and shove
Elicits tons of baby love.

THE RHINOCERUS

The rhino's skin is thick and tough,
And yet this skin is soft enough
That baby rhinos always sense,
A love enormous and intense.

THE ELEPHANT

Though mother may be short on arms,
Her skin is full of warmth and charms.
And mother's touch on baby's skin
Endears the heart that beats within.

THE CROCODILE

Here is the skin they love to touch.
It isn't soft and there isn't much,
But its contact comfort will beguile
Love from the infant crocodile.

You see, all God's chillun's got skin.

One function of the real mother, human or subhuman, and presumably of a mother surrogate, is to provide a haven of safety for the infant in times of fear and danger. The frightened or ailing child clings to its parent, not to a stranger; and this selective responsiveness in times of distress, disturbance, or danger may be used as a measure of the strength of affectional bonds. We have tested this kind of differential responsiveness by presenting to the infants in their cages, in the presence of the two mothers, various fear-producing stimuli, such as the moving toy illustrated in Figure 35. A typical response to a fear stimulus is shown in Figure 36, and the data on differential responsiveness are presented in Figure 37. It is apparent that the cloth mother

Fig. 35. Typical fear stimulus.

Fig. 36. Typical response to cloth mother in fear situations.

is highly preferred over the wire one, and this differential selectivity is enhanced by age and experience. In this situation, the variable of nursing appears to be of absolutely no importance: The infant consistently seeks the soft mother surrogate regardless of nursing condition.

Similarly, the mother or mother surrogate provides its young with a source of security, and this role or function is seen with special clarity when mother and child are in a strange situation. At the present time, we have completed tests for this relationship on four out of eight baby monkeys assigned to the dual mother-surrogate condition by introducing them for 3 minutes into the strange environment of a room measuring 6 feet × 6 feet × 6 feet (also called

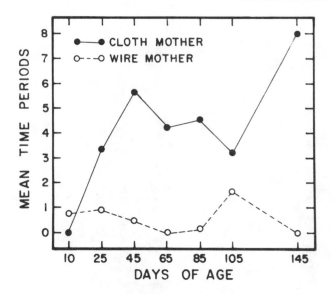

Fig. 37. Differential responsiveness in fear tests.

the "open-field test") and containing multiple stimuli known to elicit curiosity-manipulatory responses in baby monkeys. The subjects were placed in this situation twice a week for 8 weeks with no mother surrogate present during alternate sessions and the cloth mother present during the others. A cloth diaper was always available as one of the stimuli throughout all sessions. After one or two adaptation sessions, the infants always rushed to the mother surrogate when she was present and clutched her, rubbed their bodies against her, and frequently manipulated her body and face. After a few additional sessions, the infants began to use the mother surrogate as a source of security, a base of operations. As is shown in Figures 38 and 39, they would explore and manipulate a stimulus and then return to the mother before adventuring again into the strange new world. The behavior of these infants was quite different when the mother was absent from the room. Frequently they would freeze in a crouched position, as is illustrated in Figure 40. Emotionality indices such as vocalization, crouching, rocking, and sucking increased sharply, as shown in Figure 41. Total emotionality score was cut in half when the mother was present. In the absence of the mother, some of the experimental monkeys would rush to the center of the room where the mother was customarily placed and then run rapidly from object to object, screaming and crying all the while. Continuous, frantic clutching of their bodies was very common, even when not in the crouching position. These monkeys frequently contacted and clutched the cloth diaper, but this action never pacified them.

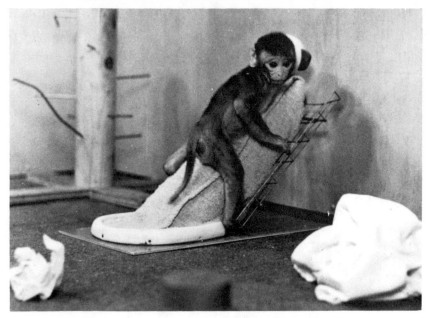

Fig. 38. Visual exploration while clinging to cloth mother.

Fig. 39. Bringing home the day's discovery.

Fig. 40. Response to wide open space in absence of surrogate mother.

The same behavior occurred in the presence of the wire mother. No difference between the cloth-mother-fed and wire-mother-fed infants was demonstrated under either condition. Four control infants never raised with a mother surrogate showed the same emotionality scores when the mother was absent as the experimental infants showed in the absence of the mother, but the controls' scores were slightly larger in the presence of the mother surrogate than in her absence.

Some years ago Robert Butler demonstrated that mature monkeys enclosed in a dimly lighted box would open and reopen a door, hour after hour, for no other reward than that of looking outside the box. We now have data indicating that neonatal monkeys show this same compulsive visual curiosity on their first test day in an adaptation of the Butler apparatus which we call the "love machine," an apparatus designed to measure love. Usually these tests are begun when the monkey is 10 days of age, but this same persistent visual exploration has been obtained in a 3-day-old monkey during the first half-hour of testing. Butler also demonstrated that rhesus monkeys show

FREE FIELD
COMPOSITE EMOTIONAL INDEX

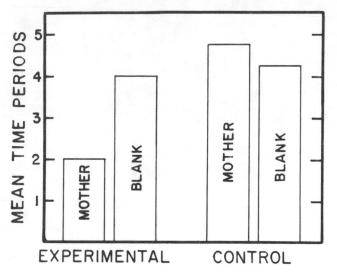

Fig. 41. Emotionality level in free open field with and without presence of cloth mother.

selectivity in rate and frequency of door-opening to stimuli of differential attractiveness in the visual field outside the box. We have utilized this principle of response selectivity by the monkey to measure strength of affectional responsiveness in our infants in the baby version of the Butler box. The test sequence involves four repetitions of a test battery in which four stimuli—cloth mother, wire mother, infant monkey, and empty box—are presented for a 30-minute period on successive days. The first four subjects in the dual mother surrogate group were given a single test sequence at 40 to 50 days of age, depending upon the availability of the apparatus, and only their data are presented. The second set of four subjects is being given repetitive tests to obtain information relating to the development of visual exploration. The apparatus is illustrated in Figure 42. The data obtained from the first four infants raised with the two mother surrogates are presented in the middle graph of Figure 43 and show approximately equal responding to the cloth mother and another infant monkey, and no greater responsiveness to the wire mother than to an empty box. Again, the results are independent of the kind of mother than lactated, cloth or wire. The same results are found for a control group raised, but not fed, on a single cloth mother; these data appear

Fig. 42. Visual exploration apparatus.

in the graph on the right. Contrariwise, the graph on the left shows no differential responsiveness to cloth and wire mothers by a second control group, which was not raised on any mother surrogate. We can be certain that not all love is blind.

The first four infant monkeys in the dual mother surrogate group were separated from their mothers between 165 and 170 days of age and tested for retention during the following 9 days and then at 30-day intervals for 6 successive months. Affectional retention as measured by the modified Butler box is given in Figure 44. In keeping with the data obtained on adult monkeys by Butler, we find a high rate of responding to any stimulus, even

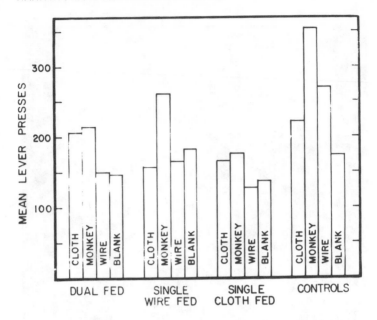

Fig. 43. Relative interest in objects visually explored.

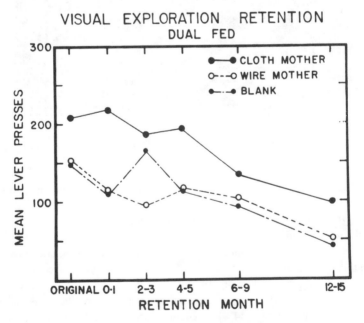

Fig. 44. Retention of exploration preferences.

the empty box. But throughout the entire 185-day retention period there is a consistent and significant difference in response frequency to the cloth mother contrasted with either the wire mother or the empty box, and no consistent difference between wire mother and empty box.

Affectional retention was also tested in the open field during the first 9 days after separation and then at 30-day intervals, and each test condition was run twice at each retention interval. The infant's behavior differed from that observed during the period preceding separation. When the cloth mother was present in the post-separation period, the babies rushed to her, climbed up, clung tightly to her, and rubbed their heads and faces against her body. After this initial embrace and reunion, they played on the mother, including biting and tearing at her cloth cover; but they rarely made any attempt to leave her during the test period, nor did they manipulate or play with the objects in the room, in contrast with their behavior before maternal separation. The only exception was the occasional monkey that left the mother surrogate momentarily, grasped the folded piece of paper (one of the standard stimuli in the field), and brought it quickly back to the mother. It appeared that deprivation had enhanced the tie to the mother and rendered the contact-comfort need so prepotent that need for the mother overwhelmed the exploratory motives during the brief, 3-minute test sessions. No change in these behaviors was observed throughout the 185-day period. When the mother was absent from the open field, the behavior of the infants was similar in the initial retention test to that during the preseparation tests; but they tended to show gradual adaptation to the open-field situation with repeated testing and, consequently, a reduction in their emotionality scores.

In the last five retention test periods, an additional test was introduced in which the surrogate mother was placed in the center of the room and covered with a clear Plexiglas box. The monkeys were initially disturbed and frustrated when their explorations and manipulations of the box failed to provide contact with the mother. However, all animals adapted to the situation rather rapidly. Soon they used the box as a place of orientation for exploratory and play behavior, made frequent contacts with the objects in the field, and very often brought these objects to the Plexiglas box. The emotionality index was slightly higher than in the condition of the available cloth mothers, but it in no way approached the emotionality level displayed when the cloth mother was absent. Obviously the infant monkeys gained emotional security by the presence of the mother even though contact was denied.

Affectional retention has also been measured by tests in which the monkey must unfasten a three-device mechanical puzzle to obtain entrance into a compartment containing the mother surrogate. All the trials are initiated by allowing the infant to go through an unlocked door, and in half the trials it finds the mother present and in half, an empty compartment. The door is then locked and a 10-minute test conducted. In tests given prior to separation from the surrogate mothers, some of the infants had solved this puzzle and others had failed. On the last test before separation, there were no differences in total manipulation under mother-present and mother-absent conditions, but striking differences exist between the two conditions throughout the post-separation test periods. Again, there is no interaction with conditions of feeding.

The overall picture obtained from surveying the retention data is unequivocal. There is little, if any, waning of responsiveness to the mother throughout this 5-month period, as indicated by any measure. It becomes perfectly obvious that this affectional bond is highly resistant to forgetting and that it can be retained for very long periods of time by relatively infrequent contact reinforcement. Later on, retention tests will be conducted at 90-day intervals, and further plans are dependent upon the results obtained. It would appear that affectional responses may show as much resistance to extinction as has been previously demonstrated for learned fears and learned pain, and such data would be in keeping with those of common human observation. The infant's responses to the mother surrogate in the fear tests, the open-field situation, and the baby Butler box, and the responses on the retention tests, cannot be described adequately with words; supplementary information can be obtained by viewing the motion picture record.

We have already described the group of four control infants that had never lived in the presence of any mother surrogate and had demonstrated no sign of affection or security in the presence of the cloth mothers introduced in test sessions. When these infants reached the age of 250 days, cubicles containing both a cloth mother and a wire mother were attached to their cages. There was no lactation in these mothers, for the monkeys were on a solid-food diet. The initial reaction of the monkeys to the alterations was one of extreme disturbance. All the infants screamed violently and made repeated attempts to escape the cage whenever the door was opened. They kept a maximum distance from the mother surrogates and exhibited a considerable amount of rocking and crouching behavior, indicative of emotionality. Our first thought was that the critical period for the development of maternally directed affection had passed and that these macaque children were doomed to live as affectional orphans. Fortunately, these behaviors continued for only 12 to 48 hours and then gradually ebbed, changing from indifference to active contact on, and exploration of, the surrogates. The home-cage behavior of these control monkeys slowly became similar to that of the animals raised with the mother surrogates from birth. Their manipulation and play on the cloth mother became progressively more vigorous, to the point of actual mutilation, particularly during the morning after the cloth mother had been given her daily change of terry covering. The control subjects were actively running to the cloth mother when frightened and had to be coaxed from her to be taken from the cage for formal testing.

Objective evidence of these changing behaviors is given in Figure 45, which plots the amount of time these infants spent on the mother surrogates. Within 10 days, mean contact time is approximately 9 hours, and this measure remains relatively constant throughout the next 30 days. Consistent with the results on the subjects reared from birth with dual mothers, these late-adopted infants spent less than 1½ hours per day in contact with the wire mothers, and this activity level was relatively constant throughout the test sessions. Although the maximum time that the control monkeys spent on the cloth mother was only about half that spent by the original dual mother-surrogate group, we cannot be sure that this discrepancy is a function of differential early experience. The control monkeys were about 3 months older when the mothers were attached to their cages than the experimental animals had been when their mothers were removed and the retention tests begun. Thus, we do not know what the amount of contact would be for a 250-day-old animal

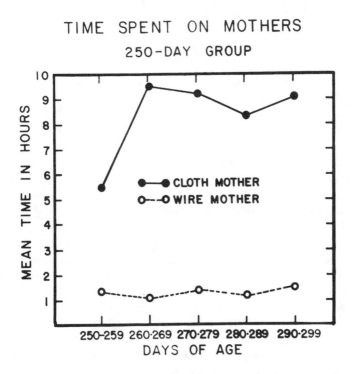

Fig. 45. Time spent on each mother by monkeys started at 250 days of age.

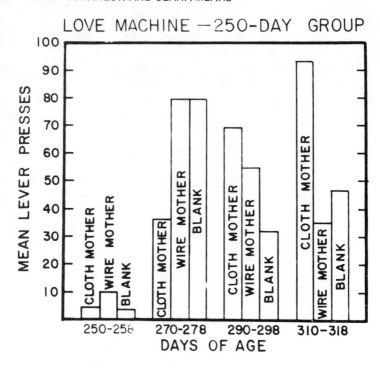

Fig. 46. Differential visual exploration of monkeys started at 250 days of age.

raised from birth with surrogate mothers. Nevertheless, the magnitude of the differences and the fact that the contact-time curves for the mothered-from-birth infants had remained constant for almost 150 days suggest that early experience with the mother is a variable of measurable importance.

The control group has also been tested for differential visual exploration after the introduction of the cloth and wire mothers; these behaviors are plotted in Figure 46. By the second test session, a high level of exploratory behavior had developed, and the responsiveness to the wire mother and the empty box is significantly greater than that to the cloth mother. This is probably not an artifact, since there is every reason to believe that the face of the cloth mother is a fear stimulus to most monkeys that have not had extensive experience with this object during the first 40 to 60 days of life. Within the third test session a sharp change in trend occurs, and the cloth mother is then more frequently viewed than the wire mother or the blank box; this trend continues during the fourth session, producing a significant preference for the cloth mother.

Before the introduction of the mother surrogate into the home-cage situation, only one of the four control monkeys had ever contacted the cloth mother in the open-field tests. In general, the surrogate mother not only gave the infants no security, but instead appeared to serve as a fear stimulus. The emotionality scores of these control subjects were slightly higher during the mother-present test sessions than during the mother-absent test sessions. These behaviors were changed radically by the fourth post-introduction test approximately 60 days later. In the absence of the cloth mothers, the emotionality index in this fourth test remains near the earlier level, but the score is reduced by half when the mother is present, a result strikingly similar to that found for infants raised with the dual mother surrogates from birth. The control infants now show increasing object exploration and play behavior, and they begin to use the mother as a base of operations, as did the infants raised from bith with the mother surrogates. However, there are still definite differences in the behavior of the two groups. The control infants do not rush directly to the mother and clutch her violently; but instead they go toward, and orient around, her, usually after an initial period during which they frequently show disturbed behavior, exploratory behavior, or both.

That the control monkeys develop affection or love for the cloth mother when she is introduced into the cage at 250 days of age cannot be questioned. There is every reason to believe, however, that this interval of delay depresses the intensity of the affectional response below that of the infant monkeys that were surrogate-mothered from birth onward. In interpreting these data, it is well to remember that the control monkeys had had continuous opportunity to observe and hear other monkeys housed in adjacent cages and that they had had limited opportunity to view and contact surrogate mothers in the test situations, even though they did not exploit the opportunities.

During the last 2 years, we have observed the behavior of two infants raised by their own mothers. Love for the real mother and love for the surrogate mother appear to be very similar. The baby macaque spends many hours a day clinging to its real mother. If away from the mother when frightened, it rushes to her and in her presence shows comfort and composure. As far as we can observe, the infant monkey's affection for the real mother is strong, but no stronger than that of the experimental monkey for the surrogate cloth mother, and the security that the infant gains from the presence of the real mother is no greater than the security it gains from a cloth surrogate. Later on, we hope to put this problem to final, definitive, experimental test. But, whether the mother is real or a cloth surrogate, there does develop a deep and abiding bond between mother and child. In one case, it may be the call of the wild and, in the other, the McCall of civilization, but in both cases there is "togetherness."

In spite of the importance of contact comfort, there is reason to believe that other variables of measurable importance will be discovered. Postural support may be such a variable, and it has been suggested that, when we build arms into the mother surrogate, 10 is the minimal number required to provide adequate child care. Rocking motion may be such a variable, and we are comparing rocking and stationary mother surrogates and inclined planes. The differential responsiveness to cloth mother and cloth-covered inclined plane suggests that clinging as well as contact is an affectional variable of importance. Sounds, particularly natural, maternal sounds, may operate as either unlearned or learned affectional variables. Visual responsiveness may be such a variable, and it is possible that some semblance of visual imprinting may develop in the neonatal monkey. There are indications that this becomes a variable of importance during the course of infancy through some maturational process.

John Bowlby (1969) has suggested that there is an affectional variable which he called "primary object following," characterized by visual and oral search of the mother's face. Our surrogate-mother-raised baby monkeys are, at first, inattentive to her face, as are human neonates to human mother faces. But by 30 days of age, ever-increasing responsiveness to the mother's face appears—whether through learning, maturation, or both—and we have reason to believe that the face becomes an object of special attention.

Our first surrogate-mother-raised baby had a mother whose head was just a ball of wood, since the baby was a month early and we had not had time to design a more esthetic head and face. This baby had contact with the blank-faced mother for 180 days and was then placed with two cloth mothers, one motionless and one rocking, both being endowed with painted, ornamented faces. To our surprise, the animal would compulsively rotate both faces 180 degrees so that it viewed only a round, smooth face and never the painted, ornamented face. Furthermore, it would do this as long as the patience of the experimenter in reorienting the faces persisted. The monkey showed no sign of fear or anxiety, but it showed unlimited persistence. Subsequently, it improved its technique, compulsively removing the heads and rolling them into its cage as fast as they were returned. We are intrigued by this observation, and we plan to examine systematically the role of the mother face in the development of infant-monkey affections. Indeed, these observations suggest the need for a series of ethological-type researches on the two-faced female.

Although we have made no attempts thus far to study the generalization of infant-macaque affection or love, the techniques which we have developed offer promise in this uncharted field. Beyond this, there are few, if any, technical difficulties in studying the affection of the actual, living mother for the child, and the techniques developed can be utilized and expanded for the analysis and developmental study of father-infant and infant-infant affection.

Since we can measure neonatal and infant affectional responses to mother surrogates, and since we know they are strong and persisting, we are in a position to assess the effects of feeding and contactual schedules; consistency and inconsistency in the mother surrogates; and early, intermediate, and late maternal deprivation. Again, we have here a family of problems of fundamental interest and theoretical importance.

If the researches completed and proposed make a contribution, I shall be grateful: But I have also given full thought to possible practical applications. The socioeconomic demands of the present and the threatened socioeconomic demands of the future have led the American woman to displace, or threaten to displace, the American man in science and industry. If this process continues, the problem of proper child-rearing practices faces us with startling clarity. It is cheering, in view of this trend, to realize that the American male is physically endowed with all the really essential equipment to compete with the American female on equal terms in one essential activity: the rearing of infants. We now know that women in the working classes are not needed in the home because of their primary mammalian capabilities: And it is possible that in the foreseeable future neonatal nursing will not be regarded as a necessity, but as a luxury—to use Veblen's term—a form of conspicuous consumption limited, perhaps, to the upper classes. But whatever course history may take, it is comforting to know that we are now in contact with the nature of love.

CHAPTER 8

THE NATURE OF LOVE– SIMPLIFIED

The cloth surrogate and its wire surrogate sibling (see Fig. 47) entered into scientific history as of 1958. The cloth surrogate was originally designed to test the relative importance of body contact in contrast to activities associated with the breast, and the results were clear beyond all expectation. Body contact was of overpowering importance by any measure taken, even contact time, as shown in Figure 48.

However, the cloth surrogate, beyond its power to measure the relative importance of a host of variables determining infant affection for the mother, exhibited another surprising trait, one of great independent usefulness. Even though the cloth mother was inanimate, it was able to impart to its infant such emotional security that the infant would, in the surrogate's presence, explore a strange situation and manipulate available physical objects (see Fig. 49) or animate objects. Manipulation of animate objects leads to play if these animate objects are age-mates, and play is the variable of primary importance in the development of normal social, sexual, and maternal functions, as described by Harlow and Harlow (1965). It is obvious that surrogate mothers, which are more docile and manipulative than real monkey mothers, have a wide range of experimental uses.

Fig. 47. Cloth and wire surrogates.

Although the original surrogates turned out to be incredibly efficient dummy mothers, they presented certain practical problems. The worst of the problems was that of cleanliness. Infant monkeys seldom soil their real mothers' bodies, though we do not know how this is achieved. However, infant monkeys soiled the bodies of the original cloth surrogates with such efficiency and enthusiasm as to present a health problem and, even worse, a financial problem resulting from laundering. Furthermore, we believed that the original cloth surrogate was too steeply angled and thereby relatively inaccessible for cuddly clinging by the neonatal monkey.

In the hope of alleviating practical problems inherent in the original cloth surrogate, we constructed a family of simplified surrogates. The simplified surrogate is mounted on a rod attached to a lead base 4 inches in diameter, angled upward at $25°$, and projected through the surrogate's body for 4 inches, so that heads may be attached if desired. The body of the simplified surrogate is only 6 inches long, 2½ inches in diameter, and stands approximately 3 inches off the ground. Figure 50 shows an original cloth surrogate and simplified surrogate placed side by side.

As can be seen in Figure 51, infants readily cling to these simplified surrogates of smaller body and decreased angle of inclination. Infant monkeys do soil the simplified surrogate, but the art and act of soiling is very greatly reduced. Terry cloth slipcovers can be made easily and relatively cheaply, alleviating, if not eliminating, laundry problems. Thus, the simplified surrogate is a far more practical dummy mother than the original cloth surrogate.

Although the original surrogate studies (Harlow, 1958; Harlow & Zimmermann, 1959) were written as if activities associated with the breast, particularly nursing, were of no importance, this is doubtlessly incorrect. There were no statistically significant differences in time spent by the babies on the lactating vs. nonlactating cloth surrogates and on the lactating vs. nonlactating wire surrogates. But the fact is that there were consistant preferences for both the cloth and the wire lactating surrogates and that these tendencies held for both the situations of time on the surrogate and the frequency of surrogate preference when the infant was exposed to a fear stimulus. Thus, if one can accept a statistically insignificant level of confidence, consistently obtained from four situations, one will properly

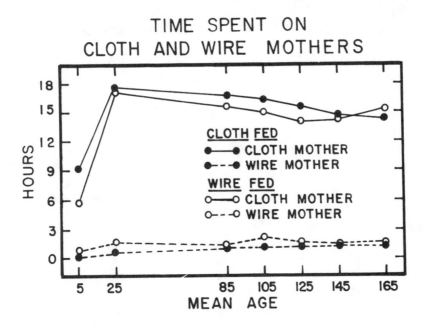

Fig. 48. Contact time to cloth and wire surrogates.

Fig. 49. Infant monkey security in presence of cloth surrogate.

conclude that nursing is a minor variable but one of more than measurable importance operating to bind the infant to the mother.

To demonstrate experimentally that activities associated with the breasts were variables of significant importance, we built two sets of differentially colored surrogates, tan and light blue; and, using a 2 × 2 Latin square design, we arranged a situation such that the surrogate of one color lactated and the other did not. As can be seen in Figure 52, the infants showed a consistent preference for the lactating surrogate when contact comfort was held constant. The importance of the lactational variable probably decreases with time. But at least we had established the hard fact that hope springs eternal in the human breast and even longer in the breast, undressed.

In the original surrogates we created an ornamental face for the cloth surrogate and a simple dog face for the wire surrogate. I was working with few available infants and against time to prepare a presidential address for the 1958 American Psychological Association Convention. On the basis of sheer intuition, I was convinced that the ornamental cloth-surrogate face would

Fig. 50. Original cloth surrogate and simplified surrogate.

Fig. 51. Infant clinging to simplified surrogate.

131

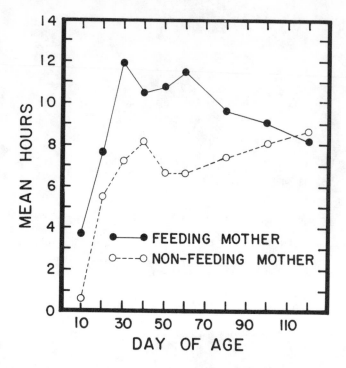

Fig. 52. Infant's preference for lactating mother.

become a stronger fear stimulus than the dog face when fear of the unfamiliar matured in the monkeys from about 70 to 110 days (Harlow & Zimmermann, 1959; Sackett, 1966). But since we wanted each surrogate to have an identifiable face and had few infants, we made no effort to balance faces by resorting to a feebleminded 2 X 2 Latin square design.

Subsequently, we have run two brief unpublished experiments. We tested four rhesus infants unfamiliar with surrogate faces at approximately 100 days of age and found that the ornamental face was a much stronger fear stimulus than the dog face. Clearly, the early enormous preference for the cloth surrogate over the wire surrogate was not a function of the differential faces. Later, we raised two infants on cloth and two on wire surrogates, counterbalancing the ornamental and dog faces. Here, the kind of face was a nonexistent variable. To a baby, all maternal faces are beautiful. A mother's face that will stop a clock will not stop an infant.

The first surrogate mother we constructed came a little late, or, phrasing it another way, her baby came a little early. Possibly her baby was illegitimate. Certainly it was her first baby. In desperation we gave the mother a face that was nothing but a round wooden ball, which displayed no trace of shame. To the baby monkey this featureless face became beautiful, and she frequently caressed it with hands and legs, beginning around 30–40 days of age. By the time the baby had reached 90 days of age, we had constructed an appropriate ornamental cloth-mother face, and we proudly mounted it on the surrogate's body. The baby took one look and screamed. She fled to the back of the cage and cringed in autistic-type posturing. After some days of terror, the infant solved the medusa-mother problem in a most ingenious manner. She revolved the face 180°. Within a week the baby resolved her unfaceable problem once and for all. She lifted the maternal head from the body, rolled it into the corner, and abandoned it. No one can blame the baby. She had lived with and loved a faceless mother, but she could not love a two-faced mother.

These data imply that an infant visually responds to the earliest version of mother he encounters, that the mother he grows accustomed to is the mother he relies upon. Subsequent changes, especially changes introduced after maturation of the fear response, elicit this response with no holds barred. Comparisons of effects of baby-sitters on human infants might be made.

We have received many questions and complaints concerning the surrogate surfaces, wire and terry cloth, used in the original studies. This mountain of mail breaks down into two general categories: that wire is aversive, and that other substances would be equally effective, if not better than, terry cloth in eliciting a clinging response.

The answer to the first matter in question is provided by observation: Wire is not an aversive stimulus to neonatal monkeys, for they spend much time climbing on the sides of their hardware-cloth cages and exploring this substance orally and tactually. A few infants have required medical treatment from protractedly pressing their faces too hard and too long against the cage sides. Obviously, however, wire does not provide contact comfort.

In an attempt to quantify preference of various materials, an exploratory study was performed in which each of four infants was presented with a choice between surrogates covered with terry cloth vs. rayon, vinyl, or rough-grade sandpaper. As shown in Figure 53, the infants demonstrated a clear preference for the cloth surrogates, and no significant preference difference between the other body surfaces.

Originally, we pointed out that rocking motion, i.e., proprioceptive stimulation, was a variable of more than statistical significance, particularly early in the infant's life, in binding the infant to the mother figure. We measured this by comparing the time the infants spent on two identical planes, one rocking and one stationary and two identical cloth surrogates, one rocking and one stationary (see Fig. 54).

To study another variable, temperature, we created some "hot mama" surrogates. We did this by inserting heating coils in the maternal bodies that

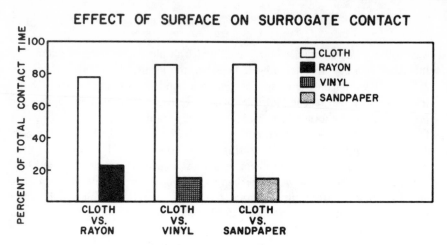

Fig. 53. Effect of surface on infant's contact with surrogate.

TIME ON MOTHERS
ROCKING VS. STATIONARY

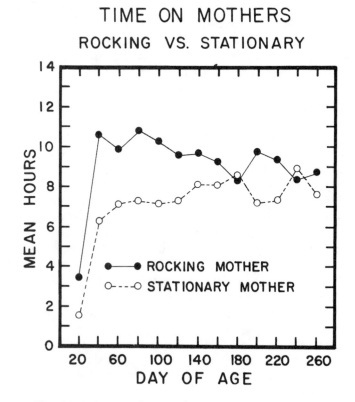

Fig. 54. Infant preference for rocking mother over stationary mother.

raised the external surrogate body surface about 10°F. In one experiment, we heated the surface of a wire surrogate and let four infant macaques choose between this heated mother and a room-temperature cloth mother. The data are presented in Figure 55. The neonatal monkeys clearly preferred the former. With increasing age this difference decreased, and at approximately 15 days the preference reversed. In a second experiment, we used two differentially colored cloth surrogates and heated one and not the other. The infants preferred the hot surrogate, but frequently contacted the room-temperature surrogate for considerable periods of time.

More recently, a series of ingenious studies on the temperature variable has been conducted by Suomi, who created hot- and cold-running surrogates by adaptation of the simplified surrogate. These results are important not only for the information obtained concerning the temperature variable, but also as an illustration of the successful experimental use of the simplified surrogate itself.

The surrogates used in these exploratory studies were modifications of the basic simplified surrogate, designed to get maximum personality out of the minimal mother. One of these surrogates was a "hot mama," exuding warmth from a conventional heating pad wrapped around the surrogate frame and

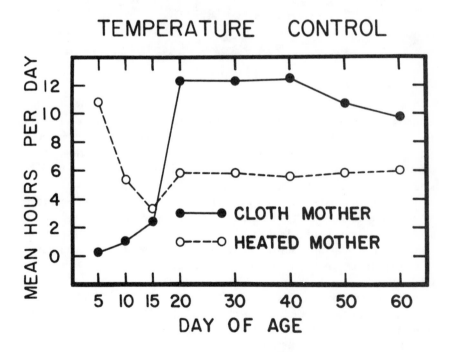

Fig. 55. Most like it hot, few like it cold.

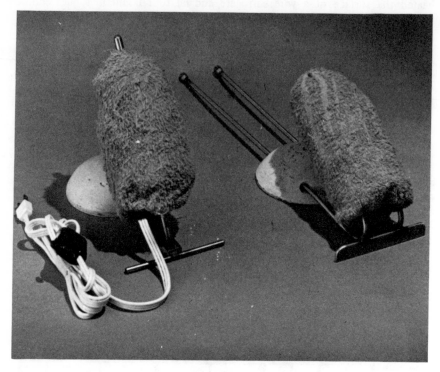

Fig. 56. Warm (left) and cold simplified surrogates.

completely covered by a terry cloth sheath. The other surrogate was a cold female; beneath the terry cloth sheath was a hollow shell within which her life fluid—cold water—was continuously circulated. The two surrogates are illustrated in Figure 56, and to the untrained observer they look remarkably similar. But looks can be deceiving, especially with females, and we felt that in these similar-looking surrogates we had really simulated the two extremes of womanhood—one with a hot body and no head, and one with a cold shoulder and no heart. Actually, this is an exaggeration, for the surface temperature of the hot surrogate was only 7°F. above room temperature, while the surface temperature of the cold surrogate was only 5°F. below room temperature.

In a preliminary study, we raised one female infant from Day 15 on the warm surrogate for a period of 4 weeks. Like all good babies, she quickly and completely became attached to her source of warmth; and during this time she exhibited not only a steadily increasing amount of surrogate contact, but also began to use the surrogate as a base for exploration (see Figs. 57). At the

Fig. 57. Infant clinging to and exploring from warm surrogate.

end of this 4-week period, we decided that our subject had become spoiled enough and so we replaced the warm surrogate with the cold version for one week. The infant noticed the switch within 2 minutes, responding by huddling in a corner and vocalizing piteously. Throughout the week of bitter maternal cold, the amount of surrogate contact fell drastically; in general, the infant avoided the surrogate in her feeding, exploratory, and sleeping behaviors. Feeling somewhat guilty, we switched surrogates once more for a week and were rewarded for our efforts by an almost immediate return to previously high levels of surrogate contact. Apparently, with heart-warming heat, our infant was capable of forgiveness, even at this tender age. At this point, we switched the two surrogates daily for a total of 2 weeks, but by this time the infant had accepted the inherent fickle nature of her mothers. On the days that her surrogate was warm, she clung tightly to its body, but on the days when the body was cold, she generally ignored it, thus providing an excellent example of naive behaviorism.

With a second infant we maintained this procedure but switched the surrogates, so that he spent four weeks with the cold surrogate, followed by one week with the warm, an additional week with the cold, and finally a two-week period in which the surrogates were switched daily. This infant became anything but attached to the cold surrogate during the initial four-week period, spending most of his time huddling in the corner of his cage and generally avoiding the surrogate in his exploratory behavior (see Figs. 58). In suceeding weeks, even with the warm surrogate, he failed to approach the levels of contact exhibited by the other infant to the cold surrogate. Apparently, being raised with a cold mother had chilled him to mothers in general, even those bearing warmth and comfort.

Two months later both infants were exposed to a severe fear stimulus in

Fig. 58. Typical infant reactions to cold simplified surrogate.

the presence of a room-temperature simplified surrogate. The warm-mother infant responded to this stimulus by running to the surrogate and clinging for dear life. The cold-mother infant responded by running the other way and seeking security in a corner of the cage. We seriously doubt that this behavioral difference can be attributed to the sex difference of our subjects. Rather, this demonstration warmed our hopes and chilled our doubts that temperature may be a variable of importance. More specifically, it suggested that a simple linear model may not be adequate to describe the effects of temperature differences of surrogates on infant attachment. It is clear that warmth is a variable of major importance, particularly in the neonate, and we hazard the guess that elevated temperature is a variable of importance in the operation of all the affectional systems: maternal, mother-infant, possibly agemate, heterosexual, and even paternal.

We simplified the surrogate mother further for studies in which its only function is that of providing early social support and security to infants. This supersimplified surrogate is merely a board 1½ inches in diameter and 10 inches long with a scooped-out, concave trough having a maximal depth of ¾ inch. The supersimplified surrogate has an angular deviation from the base of less than 15°, though this angle can be increased by the experimenter at will. The standard cover for this supremely simple surrogate mother is a size 11, cotton athletic sock, though covers of various qualities, rayon, vinyl (which we call the "linoleum lover"), and sandpaper, have been used for experimental purposes.

> Linoleum lover, with you I am through
> The course of smooth love never runs true.

This supersimplified mother is designed to attract and elicit clinging responses from the infant during the first 15 days of the infant's life.

We have designed, but not yet tested, a swinging mother that will dangle from a frame about 2 inches off the floor and have a convex, terry cloth or cotton body surface. Observations of real macaque neonates and mothers indicate that the infant, not the mother, is the primary attachment object even when the mother locomotes, and that this swinging mother may also elicit infantile clasp and impart infant security very early in life. There is nothing original in this day and age about a swinger becoming a mother, and the only new angle, if any, is a mother becoming a swinger.

Additional findings, such as the discovery that 6-month social isolates will learn to cling to a heated simplified surrogate, and that the presence of a surrogate reduces clinging among infant-infant pairs, have substantiated use of the surrogate beyond experiments for its own sake. At present, the heated simplified surrogate is being utilized as a standard apparatus in studies as varied as reaction to fear, rehabilitation of social isolates, and development of play. To date, additional research utilizing the cold version of the simplified surrogate has been far more limited, possibly because unused water faucets are harder to obtain than empty electrical outlets. But this represents a methodological, not a theoretical problem, and without doubt solutions will soon be forthcoming.

It is obvious that the surrogate mother at this point is not merely a historical showpiece. Unlike the proverbial old soldier, it is far from fading away. Instead, as in the past, it continues to foster not only new infants but new ideas.

CHAPTER 9

THE POWER AND PASSION
OF PLAY

Behaviors sheltering under the ubiquitous umbrella called play are as diverse as the theories proposed and propounded to account for their existence. The area of behavior termed play has not been historically ignored nor neglected yet remains, scientifically, an almost unmarked maze which has defied systematic exploration. The only general consensus on the subject of play is an acceptance of the statement made by Hurlock (1934) 40 years ago that there is no agreement among writers about play. The literature is permeated with normative studies which catalog activity preferences, toy preferences, and preferences for every game from tiddley-winks to camber casting. Many of the studies of play reflect the biased assumption that play must have a personal or utilitarian function. For example, Spencer (1873) conceived of play as a release of surplus energy, whereas Groos (1901) viewed infant play as a classroom for direct training for future adult activity. Many other investigators have taken the Freudian position that emotional conflicts and aggressions are alleviated or resolved through the vehicle of imaginative play, such as play with dolls, with finger paints, or even with gooey dough.

141

There has been far greater agreement as to what constitutes the form and function of play among members of the nonhuman than the human primate species. This is probably the case because scientists are not inclined to postulate ulterior motives for nonhuman animal play. Research with these primates has been devoted to social play to such an extent that references to play among subhuman primates are practically synonymous to references to social play, i.e., play with playmates.

The play of monkeys in the monkey house at zoos has produced pleasure in countless children and admiring adults throughout the years. Their agile antics make these animals look like lively, leaping caricatures of furry little men with tails, and have enriched the human language with phrases like monkeyshines and monkey business or monkeying around. However, of all monkey behaviors, play has probably been one of the most informative in contributing information about the development of social and sexual roles.

In his study of the third affectional system, peer or agemate love and affection, Harlow searched for antecedent mechanisms responsible for the development and maintenance of positive peer interactions: Harlow concluded that play was the variable of primary and underlying importance. The systematic study of play, especially between the fourth and eighth months of life, has made possible the identification of variables vital to the emergence of adequate adult heterosexual behaviors and the creation of competent, contributing members of monkey society.

Early maternal-infant bonds inhibit the emergence of play until such a time as the baby is physically coordinated enough to maneuver under its own power and mature enough to be aware of some pitfalls of the public domain away from mother.

Another extremely important variable is the conveyance through the mother to the child of a sense of security and trust which creates self-confidence within the infant to face the future of the outer world (Fig. 59). As the baby initiates the first forays away from mother, fear of the strange and unknown are foremost and any sudden or intense stimulus sends him hurtling back to rub new contact and comfort from mother's brave, bounteous and beautiful body. Maternal contact eliminates the fear and each experience adds additional confidence. The mechanism magnificently mirrors the experimental extinction of fear in the laboratory conditioning experiments since Pavlov. In addition to contact comfort, all maternal ministrations during early mother-baby behavior beginnings—warmth, rocking, nursing, and protection—contribute to the confidence build-up.

In all probability the mother's behavior also facilitates play because the infant is learning social responsiveness in interactions with the mother. As this period progresses, the baby monkey can be observed watching the mother for approval or disapproval and for information concerning its own behavior. In

Fig. 59. The cloth mother offers security for early play.

early monkey development, mother-raised infants have been experimentally shown in fear situations to develop positive threat reactions which reduce the ineffective, maladaptive behaviors produced by the same fear stimuli with rhesus infants raised without their natural mothers (Novak, 1974). Scientific literature of the second half of the 20th century is replete with studies of the significance of early social communication between the human mother and baby through eye-contact, smiling, laughing, and early babbling.

During the second month of life, the infant monkey spends an ever-increasing amount of time away from the maternal circle, investigating both animate and inanimate objects. The monkey might approach a confrere with a series of small hops and tentative touches, which were not reciprocated because the animal touched was only reached physically and not emotionally. However, by the third month, these sorties gain salience and interaction ensues with mutual chasing, leaping, and incipient wrestling and baby-biting in sham fashion.

Between 4 and 8 months, the monkey's life is dominated by play activity, primarily of two forms described as rough-and-tumble and approach-avoidance. The development of these two patterns is almost co-extensive, with more individual than developmental differences. Rough-and-tumble play is best described as sham fighting or wrestling with active physical contact and rolling and sham biting as an integral part. Blurton-Jones (1967) found similar forms of play among English nursery school children. He forcefully stated that patterns of play almost identical to the monkey rough-and-tumble activity occur and are clearly definable among human children. Miraculously, no one, monkey or child, is hurt. Rough-and-tumble play predominates among male monkeys. Among pre-school children, boys tend to play outdoors consistently more than girls and require much more space, which is a very good idea with rough-and-tumble taking place (Harper & Sanders, 1975). The second pattern, approach-avoidance play, is probably more characteristic of the female. This play resembles the human game of tag and is represented in the many children's games involving chasing and avoidance of being caught.

Sexual posturing begins to appear as early as 4 months of monkey age. Mounting and thrusting, precursors of male sex, increase in frequency, proficiency, and enthusiasm and eventually become highly differentiated in form from their origins in play activity. Female posturing, also developed through play patterns, emerges from postural passivity to postural present. As a matter of fact, the postural presentation is at first a behavior of both sexes but becomes the primary prerogative of the female.

The mother monkey influences the development of play among peers in ways additional to the creation of self-confidence of the offspring. With human children, as well as monkey infants, play is petrified in strange and new situations. The presence of the human mother relieves the frigid and frozen atmosphere by reducing the total strangeness (Passman & Weisbey, 1975). The importance of contact comfort is also emphasized by Passman and Weisbey since children with strong security blanket attachments find the same relief from the presence of the security blanket. Monkey mothers, who in the first 2 months are almost oversolicitous of the babies' protection, relax their surveillance and actually urge the children to join peers in play with a playful push when the time is ripe. Once the infant revels in rollicking with its playmates, the mother's urging becomes unnecessary and more and more time is spent in play.

The importance of play in forming socially successful roles in sex and society is most clearly illustrated by monkeys who have been deprived of peer

Fig. 60. Raised without peer play, the male monkey sex life proves to be at cross purposes with reality.

play in their formative years. Infants reared with their mothers but without the opportunity for interaction with their own agemates for the first 8 months of life are hyperaggressive when finally exposed to peers. Instead of normal physical play contacts, they try to bite and physically attack. Like human mama's boys, they are not additions to any social group. Female monkeys completely isolated from mothers and agemates during early life resisted all approaches to sexual intercourse and male monkeys with the same background were sexually blank and banal (Fig. 60). Monkeys denied the chance to play at sex are seldom proficient at sexual play. Their hearts are the only anatomical part in the right place at the right time. During the isolation period, these monkeys of both sexes developed self-directed physical behavior, such as self-clasping, rocking, and self-biting. When placed in groups, no play behavior appeared, but, rather, the same maladaptive self-directed abnormal behaviors persisted.

Play repertoires of young monkeys contain the origins of most adult social behaviors. Patterns of social grooming, dominance, aggression, and sex are clearly evident in monkey play activity, though not at competent adult levels. At first clumsy to the point of being ridiculous, months and even years of practice at play produce the adult product.

Harlow has searched for antecedents to these complex social play behaviors. Exhaustive research, both prior and subsequent to study of peer play, has been conducted on curiosity and manipulative activity of the rhesus monkey. Monkeys, both young and old, readily explore and manipulate novel objects in the absence of any explicit external reward. Initially, monkey infants are equally curious toward social and nonsocial objects, but they soon come to prefer animate playmates to inanimate objects (Suomi & Harlow, 1971). Harlow formulated the hypothesis that unlearned curiosity and manipulation were the antecedents for social play. Both curiosity-exploration and play behaviors develop along similar maturational courses. Social play progressively burgeons after curiosity is well under way, and each successive stage of curiosity is followed by new social developments.

Although the positive relationship between curiosity and manipulation and play cannot be questioned, it is doubtful if curiosity and exploration are the direct and immediate variables leading to social play. This obvious connection puzzled Harlow, but he found no solution until his mental marriage to Mears.

Mears had been interested in the role of physical activity and of athletic games in developing and maintaining positive social adjustment. Most of the studies of children's play were conducted in the physically circumspect environment of the inside rooms of nursery schools. Very few offered the freedom and range of outdoor activities. A child may go into the house to play with dolls and trains, to play records, or to play store. A child will go out to play, for the sake of play itself, out where there is room to be free, to run, to leap, to hop, and to jump, or, in the words of Stevenson:

> "How do you like to go up in a swing,
> Up in the air so blue?
> Oh, I do think it the pleasantest thing
> Ever a child can do!"

Mears has recently postulated that the fundamental play form, primary and basic to social play, is centered around a group of behaviors appropriately termed self-motion play. This play form is not manipulation or motion of other objects by the self, but rather motion of the self as a reinforcer in and of its own right. As such, self-motion play can be differentiated from both traditional locomotor activity and the above forms of social play. Apparatus may be involved but is not obligatory. Monkey self-motion play is illustrated in Figure 61. Human self-motion play takes place primarily outdoors. When it

Fig. 61. Self-motion play.

takes place indoors, parents protest. It may be either solitary or social. A rarely used term, *peragration* (motion through space), provides a perfect description. There was some temptation to use the phrase *activity play*, but several scientific investigators have already applied this phrase to include the manual manipulation of objects, an act which may occur in the course of self-motion play but is not exclusively self-motion play itself. Kinesthetic sensations are evident in personal peragrations but, again, precendents might cause confusion. For example, Külka, Fry, and Goldstein (1960) hypothesized kinesthetic needs in infancy, with motility the modality of expression. They used "kinesthetic" to refer to sensations from light, touch, pressure, temperature, viscera, and all their central representations. In self-motion play there are kinesthetic sensations, but only in the customary scientific sense—sensations from the muscles, tendons, and joints—and self-motion play is far more than these sensations. It is the behaviors themselves. To reduce play to sensations is reduction to absurdity.

The existence of self-motion play has long been recognized by many

Fig. 62. Self-motion playroom.

scientific investigators, in fact if not in name, but no one has realized its significance to the child nor its probably long-term influence on development through adolescence and even into old age. Endless examples of such human activity extend from rocking in the cradle to rock and roll, and then on to the rocking chair as one approaches the Rock of Ages. The indoor baby swing moves outdoors to become the swing or tire under the old oak tree; next the ferris wheel or roller coaster beckon and, finally, one day a disgruntled dad unwinds in the hammock or on the front porch swing. Roller skating, ice skating, swimming, diving, dancing, water and snow skiing, the rocking horse, horseback riding—all are encompassed by self-motion play. The list lengthens indefinitely, concluding with the parachute for the brave and the merry-go-round for the meek and mousy.

    In the world of subhuman primates the story is the same. Harlow and Zimmermann (1958) experimentally demonstrated the preference of the infant rhesus for the surrogate mother who rocked over the one who only stood ungaited. Köhler's chimpanzees polevaulted with pleasure and without bribes of bananas or bonbons. Every primatologist, or any and every child who has visited the zoo, has witnessed the monkeys which fly through the air with the greatest of ease, rivaling the famed man on the flying trapeze. The behaviors of motion play become less stereotyped the higher the animal's position in the phyletic scale. "Some fishes periodically leap above the water's surface, birds indulge in elaborate aerial maneuvers, colts gallop, puppies race, and kittens scamper," according to Beach (1945). For countless years, self-motion play in many animals has been described in detail and ignored in intent, even if unintentionally.

Research of Mears and Harlow (1975) was designed to test the existence of self-motion play as an identifiable entity and to investigate the developmental relationships between social play, curiosity, exploration, and self-motion play. In order to avoid the traumatic effects of mother-infant separation during social testing with their peers, the eight rhesus babies were separated from the mothers at birth and, after a period in the laboratory nursery, raised with surrogate mothers who subsequently were with the infants in their home cages and in the specially designed playroom where testing took place.

The playroom was equipped in such a way as to encourage whatever physical pattern of activity the infants chose, be it jolly jumping or passive, peaceful, solitary or social interlude. Platforms of varied heights were placed within leaping distance of each other, with ladders of appropriate heights for escalation. A revolving wheel and ladder, a long, high bar and rings swinging from chains supplied other means of self-motion play. Oral and tactile manipulation and exploration were encouraged by rubber strips, varied toys, a ball and chain, and an interlocking puzzle. Stationary low platforms provided the equivalent of park benches (Fig. 62).

From the ages of 3 to 6 months, a complete recording of behaviors, with the exception of eating, drinking, and elimination, was made by experienced testers for 10 minutes of each of 5 days per week. These behavioral categories (condensed) are summarized and operationally defined below:

> (1) Self-motion play (apparatus, nonsocial):
>> Swinging, jumping, leaping on or from apparatus, running, acrobatics, accelerated or general bodily motion through space and involving apparatus.
> (2) Self-motion play (no apparatus, nonsocial):
>> Same as (1) but without apparatus.
> (3) Self-motion play (apparatus, social):
>> Same as (1) but with monkey interaction.
> (4) Self-motion play (no apparatus, social):
>> Same as (2) but with monkey interaction.
> (5) Self-motion play (rough-and-tumble):
>> Social play, vigorous physical contact, rolling, wrestling, sham fighting.
> (6) Locomotion:
>> Ambulatory activity not otherwise delineated.

(7) Tactile-oral exploration:
Active manual or oral contact with inanimate environmental objects.
(8) Abnormal, maladaptive behaviors:
Separately recorded were the following behaviors which, if persistent, indicate abnormality: rocking-huddling, self-mouth, self-clasp, ventral cling, surrogate contact, and infantile sex.

The primary finding lay in the clear emergence and development of a variety of forms of self-motion play. As can clearly be seen in Table 1, the first three of the five predefined forms of self-motion play showed significant developmental increases over the 12 weeks and were the only behaviors in the entire repertoire to show significant increases in frequency.

Figure 63 plots the development of self-motion play involving the apparatus in the playroom as opposed to that occurring without the use of any apparatus. The adequate apparatus in the special playroom obviously aided and abetted acts of peragration. However, self-motion play did not depend entirely upon props for its importance. Even without the enticement of the whirling wheel and the swinging rings, the infant rhesus subjects spend considerable time simply running, chasing, doing flips in the air, and tumbling on the floor.

Self-motion play, as an entity, achieved far higher frequencies of occurrence than did either locomotion or exploration in the time span covered in this study. Lest even the sophisticated observer confuse the drunken-sailor

Table 1. Time block effects during playroom testing

Category	df	Mean square	F	p	Direction
Self-motion play (apparatus, nonsocial)	11/44	320.419	6.53	$p < 0.0005$	Increasing
Self-motion play (no apparatus, nonsocial)	11/44	61.988	4.26	$p < 0.0005$	Increasing
Self-motion play (apparatus, social)	11/44	67.969	10.35	$p < 0.0005$	Increasing
Self-motion play (no apparatus, social)	11/44	6.625	1.15	NS	
Self-motion play (rough-and-tumble)	11/44	6.564	1.69	NS	
Locomotion	11/44	47.805	1.09	NS	
Environmental exploration	11/44	21.473	1.47	NS	

NS, not significant.

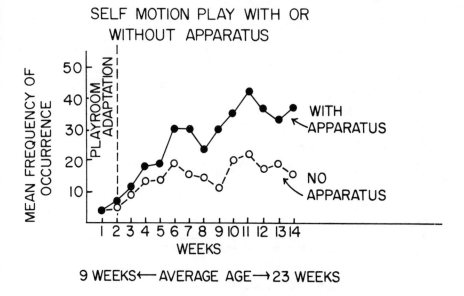

Fig. 63. Self-motion play with or without apparatus.

gait of the neonatal rhesus with self-motion play, all ambulatory behavior from one location to another, walking or climbing, was scored as locomotion unless it involved increased acceleration or greater complexity. As Figure 64 shows, the 9-week-old infants were ambulating almost three times as frequently as they were engaging in self-motion play, but locomotion rapidly reached a plateau. Self-motion play was just beginning to burgeon. By the end of the experiment, the relative frequencies of the two activities were reversed.

That exploration was not the sustaining or prime factor in self-motion play behaviors is obvious from Figure 65. There is no doubt of the importance and priority of curiosity-exploration in the initiation of all new behaviors. It is the sine qua non in the origin of different and varied activities, permitting the strange to become familiar. Once the strangeness of new self-motion play behavior disappeared, however, curiosity-exploration continued at an even pace while self-motion play rose rapidly to a frequency six times as great.

A comparison of individual and social self-motion play (Fig. 66) indicates that individual rather than social peragrations holds the preferred role in the first 6 months of rhesus life. Self-motion play can obviously be social as well as solitary, but it develops earlier and to a greater extent as an experience of the solitary animal. The present study concluded when the rhesus infants were the very age, 23 weeks, at which social contact and noncontact play begin to

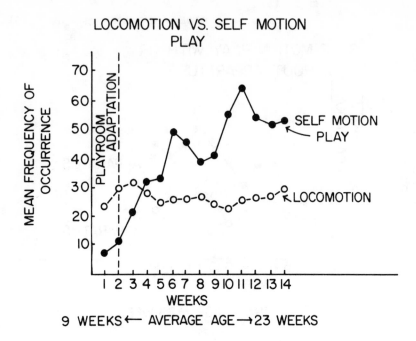

Fig. 64. Locomotion vs. self-motion play.

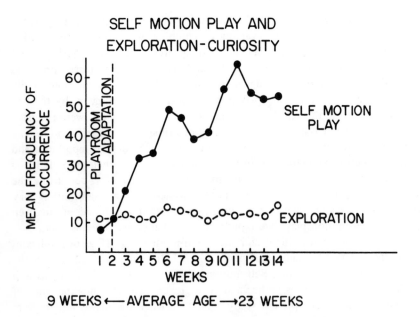

Fig. 65. Self-motion play and exploration-curiosity.

152

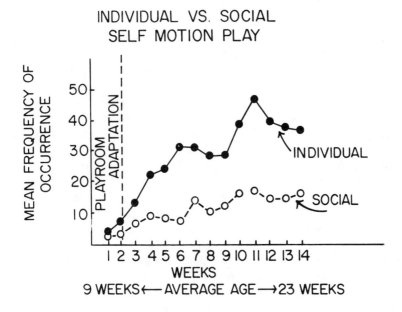

Fig. 66. Individual play predominates.

spiral (Rosevear, 1970). Supporting Rosevear's data, in this study social rough-and-tumble play did not achieve the levels shown by the other forms of self-motion play. Social self-motion play in general develops along with the stronger solitary source. It is interesting to note that well-equipped playgrounds reduce both social play and social conflicts among human children (Johnson, 1971).

In his research on nursery school children, Blurton Jones (1967) included in rough-and-tumble play more than the rough fake fighting, wrestling, and tumbling which are the primary factors in the play as described in the Wisconsin researches. Added to these behaviors were additional peragrations, running, chasing, fleeing, and jumping up and down. Above and beyond these behaviors was one of even greater importance, the laughter which accompanied all of the aspects of this play. Blurton Jones perceived the pure enjoyment of self-motion play in some of its guises and also suspected that there might be some far-reaching behavioral implications for the human child. He mentioned that some of the children new to the school did not immediately share the frolicking fun and that some children never did learn how to join in the jostle just for fun. What kind of adults, he mused, would these children become?

That self-motion play may contribute to the development of positive personality characteristics is suggested also by additional data gleaned. The recorded data on the abnormal behaviors of these eight monkeys were compared with data from sex-matched pairs of the same ages. Two behaviors indicative of psychopathology among both monkey and human infants are excessive rock-and-huddle and ventral cling. The eight rhesus with the opportunity for playroom peragration consistently showed much lower levels of both rock-and-huddle and ventral cling. Within limits, it is possible that self-motion play may act as a silent but subtle type of psychotherapy.

Recently, the wealth of information gleaned experimentally about the development of the human infant during the first 2 years of life has led Lewis (1976) to forcibly and convincingly postulate the development of a concept of competence during the first 2 years of human life. This concept develops along with the concept of self (Lewis & Brooks, 1978) which follows the emergence of recognition of people other than self. First of all, the baby separates mother from others, then strangers from mothers, and then, also during the first year of life, peers from adults. Thus, gradually, the concept of self is being established as being different from other people who are different from each other. By 2 years of age, the child verbally indicates the recognition of self by recognizing "mine" as separate from "yours" and "me" from "you."

Using the 4 to 1 ratio in comparing the developmental ages of the rhesus and the human infant, the rhesus is speedily engaging in higher and higher levels of self-motion play from the age of 3 months on, or from 1 year of human calendar age. Peragration is developing concurrently with the concepts of self and of competence. Of course, the infants, both human and monkey, have been enjoying self-motion ever since rocking along with mother, long before human adults have affixed the label of play to the behavior.

A subsequent experiment has further delineated and added to knowledge of both the prowess and powers possible in peragration (Mears, 1978). The eight rhesus infants of the first investigation, now just 6 months older than when first initiated into the playroom, were divided into two sex-matched sets of four each. One group of four continued playing together in the elegantly equipped playroom to which they were by now well accustomed. For the other four, the experimental group, every moveable piece of motion enhancing apparatus was removed completely from the playroom. The long iron bar was raised to the ceiling, as close as it could be placed, with the chains and erstwhile swinging rings wound tightly around it. The long bar left but one trace of its former position. About two-thirds of the distance, or 140 cm., up one side wall was left a tiny, projecting metal support, 1 X 1½ cm. large, on which one end of the long bar had rested. The only equipment remaining were the stationary platforms, originally placed well out of leaping distance of each other. Manipulatory toys and puzzles were slightly chewed but still to be used, and they also remained.

Peragration which needed to be pampered into playful heights with playroom apparatus obviously disappeared from the repertoire of the behavior, but all was not playfully quiet on the experimental front. Self-motion play, with absolutely no prefabricated props, no swinging rings, no high leaping platforms, no long, high bar on which to run and from which to jump, no ladders or revolving wheels, self-motion play sans apparatus, reached new levels of frequency as if it compensate for the trick played on the rhesus subjects. Their peragration without apparatus was significantly higher than that of the smugly swinging control group in the equipped playroom, and also significantly higher than their own similar play when all eight played in the playroom.

New peragration behaviors without apparatus were added to those present before. The rhesus infants created motion apparatus where none had existed. The walls became vertical jumping platforms, and the stationary playforms moved figuratively but also operationally closer to each other, now within broad jump reach. New jumping records were set and two new stars were born.

The birth of the stars was deliberate and prolonged, but without pain. One male monkey was entranced by the sight of the long bar nestled close to the ceiling. For days his gaze was fixed upon the bar and then he proceeded to practice leaps against the wall, each leap a little higher than those preceding. After seemingly hopeless but ever higher leaps, one day he caught and held onto the tiny metal know high up on the side wall. He reached out one arm, pushing against the wall and next, flattening himself and using corner walls for leverage, he literally climbed the wall. His gaze never left the long bar with the ring protruding stiffly. He maneuvered himself within reach, pulled himself up, and stood looking down on the lowly mortals below.

Even though the long bar and rings did not provide the same swinging delight as in former days, the simian fly act was repeated over days and had not gone unnoticed by the monkey confreres. The male monkey's female playmate followed, step by step, the same process and, after several days of attempts at social climbing, she, too, reached her goal and together the playmates looked down, lord and lady of all surveyed.

The concept of competence and the mastery mechanisms just described are practically inseparable. The most important distinction is that the mastery must succeed in order to help create the concept of competence. Self-motion play offers invaluable, self-reinforcing opportunities (Fig. 67) for developing competence, but the human caretaker, as in so many other aspects of developing behavior, must assume much more responsibility than the monkey mother in order to insure positive reinforcement. Without an alert adult, the

Fig. 67. No windmill whirling is needed to bring the smile to the face of the running child.

joyful first step may prove to be just a baneful bump on baby's knee or the first playground swinging too sudden and intense, a stimulus for fear and not fun.

It is our belief that in primates—monkeys and men—self-motion play begins shortly after birth (Fig. 68), as ambulatory capability matures, and continues throughout life, in changing manifestations, to contribute its integrative components and continuities to individual and social development. The data from the research with monkeys presented in this chapter show that self-motion play, during the early months of life, exists as an entity of more force and frequency than other forms of play and locomotor activity. The

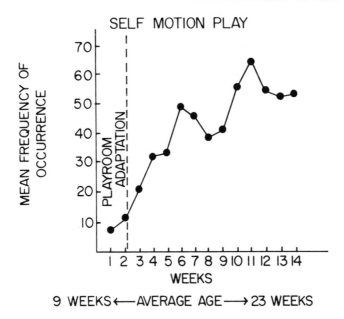

Fig. 68. Self-motion play spirals upward in early months of rhesus life.

only form of social play to increase significantly during this research was social play based on the use of apparatus, a fact which suggests the role this form of play may take in the development of social play. Self-motion play is a basic form from which other play patterns evolve and from which certain pleasures associated with perception of motion predominate. We believe that it is basically a complex, unlearned, unconditioned motor response, insofar as any primate behaviors are completely unlearned. Its power is achieved by two overriding qualities. The first is that it is primary—it emerges chronologically prior to other forms of play. Of equal importance is the persistence of self-motion play. It continues throughout developmental periods, thus providing a foundation for perfection of increasingly complex behaviors, such as self-confidence and competence. In human adults peragration achieves an apex in the precision of the professional athlete's movements or in the cultural culmination, the ballet.*

*For further experimental data and details the reader is referred to the following reference from which excerpts have been made with the approval of the source: Mears, C. E. and Harlow, H. F. Play, Early and Eternal. *Proceedings of the National Academy of Sciences,* 1975, **72**, 1878.

PART III

THE LOVES OF LIFE

CHAPTER 10

THE LINKAGE OF LOVES

For many years we have conducted experiments at the University of Wisconsin Primate Laboratory. We have studied subhuman primates in the hope of finding facts hitherto unknown about human beings. We have experimented with monkeys because there is enormous similarity between man and monkeys, though we accept the fact that man is more sentient and subtle, especially woman-man.

Now, for many years we have studied love. We have discovered many things and quite a few of them are about love. One of the facts about which we are quite sure is that, contrary to the traditional attitudes of poets and philosophers and even pedants, love is not a simple, single, unitary behavior or process.

Love may make the world go round, but there is seldom, possibly never, enough love to go around with it, even though there are five different kinds of love that all or most human beings commonly experience. These five kinds are identifiable, demonstrable, active love systems. It would, therefore, be impossible to have one specific spark to generate one generalized behavior called love.

Our study of love among the monkeys was conducted in laboratories, one laboratory after another. To many investigators a laboratory is mainly a concrete dungeon filled with abject and apathetic animals and dispirited experimenters. Perhaps many laboratories are merely reflected images of their senior scientists. This is one way to construct and create a laboratory, particularly if your interest is in achieving factual and fatuous data that are only statistically significant. If one searches for data that are socially significant as well as statistically significant, a laboratory should be more than a concrete dungeon. It should be an environment that affords its nonhuman individuals the social opportunities of the wild but with complete and perfect predator protection. The social environment of the laboratory should stimulate all of its inhabitants, including the experimenter, to achieve their full intellectual capabilities, even if this goal has not yet been achieved for people in our United States.

With the cooperation of the cloth surrogate mother, we confirmed the presence of multiple variables and the nature of these variables in the first two love systems, the infant's love for the mother with its firm and fine foundations in contact comfort (Fig. 69) and further support from the milk of maternal kindness and the warmth of human nature. The second, that of maternal love, has been universally accepted for as long as poets have offered their paeans of praise, and even more so since Mother's Day was declared by the florists. This acceptance is fortunate for the primate population whose very existence depends upon the maternal blessings bestowed upon the babies.

The third affectional system is one which has been relatively neglected in spite of the fact that it is of predominant importance for subsequent social development and the creation and conception of sex and procreating behaviors. This third love system is the agemate or peer affectional system. It is the love of babies for babies, infants for infants, and preadolescents for preadolescents. It is the love of friends for friends and as such is probably the longest lasting and most life pervasive of any of the five presently identified and accepted love systems.

The primary mechanisms binding the immature social members together are those of play, an area about which unlimited words have been written and meaningful messages are rare. The feeling of affection between agemates may appear as soon as they see eye to eye or exchange toys through force or favor, but play adds power to the passive passion.

The remaining two love systems are the heterosexual and the paternal, the love of the sexually mature woman and man for each other and the love of the father for his family, be it limited by blood or extended by cultural customs.

Let us consider paternal love next, lest it be said that father is always the last to be considered. No matter what adjustments have been made through

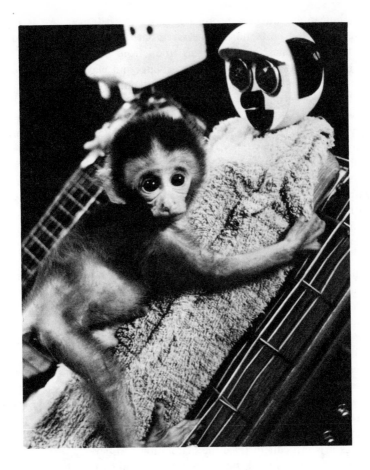

Fig. 69. Security and trust through contact comfort.

women's suffrage, no matter what alterations are being made through matriarchal societies or through equal rights amendments, maternal and paternal love shall maintain some of their discreteness unless transplants can be completely inclusive. If you transplant the entire gamut of feminine physiological functions to the male and the male functions to the female, what would you have then, except a male and a female?

Paternal love is intricately bound to his immediate or extended social group, to his family and his country or even to his country club, be he a human primate, and to his family or to his troop, be he a monkey primate. Fatherly affection may be as flexible as the social groups dictate or indicate.

Whether human or subhuman primate, however, paternal love is primarily concerned with the protection of his loved ones. The monkey male adult will protect the geographical borders of his troop from inimical strangers and the smaller infants of the group from larger siblings and playmates. If need be, he will protect the babies from an overly punitive mother.

The fifth love is the most difficult of human loves to describe and yet is probably the love which the single word, "love," usually implies when used in everyday and every year conversation. There is a suspicion that the word may, in many minds, be synonymous with sexual attraction or even with compatible, mutual sexual happiness.

And yet this same love between sexually mature male and female has designated the romantic love which thrived primarily on chivalry, the songs of the sentimental troubadors, and the concept of the sweet girl graduate.

In still other days and ages the word properly referred to marital love alone, even should the couple involved not really feel or appear to be "in love."

To put love fully in its place we must first see love in relation to the emotional systems in human and subhuman primates. In the early part of the 20th century, the emotions became instinctively entwined in the controversy over the relative importance of nature over nurture or, in other terms, of learning over heredity. Emotions became involved with instincts when psychologists were trying to escape from the tangled problem of the innateness of instincts. Instead they became embroiled in the controversy of whether or not emotions were inherited—or learned.

It has long been recognized that many aspects of emotions either can be or are learned. But while learning was usurping the center of the stage, the importance of emotional sequencing and maturation was obscured.

Whereas it is difficult to trace the maturation and development of emotions in the human child, this difficulty results not only from the complexity of human emotions but partially from inadequate experimentation and exploration. It is relatively easy to trace the maturation of emotions in the rhesus monkey and there is no reason to believe that the developmental processes of maturation differ significantly in the two species. There are, of course, differences in the absolute time of appearance of any emotion in man and monkey, but we have found the order of maturation to be very similar.

Insofar as the pleasant emotions are concerned, maternal and paternal love as well as love of the infant for the parents precede the time of appearance of any and all of the unpleasant emotions. Fear, as described by Watson (1924), is reflex fear, emotion in its most simplistic form, the startle response to sudden loss of support or to sudden loud noise. The fears really fearful and behaviorally disruptive may be as complex as infant and maternal love, but start much later. For instance, fear of strangers and strange places appears around the ninth month of infancy, albeit with a range from the sixth to the eleventh month. In monkey maturation the time of appearance of fear of strangers centers around the age of 90 days. This factual fear appears first after the earliest two loves have become well established, and the maturation of aggression follows the first loves even later.

The pleasant emotion of agemate or peer love probably precedes the unpleasant emotion of aggression even though a large part of the developmental course of peer love overlaps temporally. By the time heterosexual love appears, emotions of fear and aggression have become an integral part of the repertoire of the primate. The primate by then has, however, had ample time to develop through love and learning the inhibition or the amelioration of antisocial fears and antisocial aggressions.

Freud has received credit for the discovery of the importance of childhood and especially the importance of the very first few years of life. Many psychologists have felt dissatisfied with Freud's foremost facts on the nature of the maternal bonds which account for the importance of infancy, but still the suspicion of the importance has persisted.

It could not be that the petite infant was considered to be actively participating in the activities of the human world and offering its childish contributions since, from 1930 to 1966, there were still some professional pronouncements which described the baby as bereft of any cortical competence, a helpless, hapless human neonate with only reflex responses, sans sensory discrimination and practically unconscious (Stone, Smith, and Murphy, 1973).

The tiny, tiny baby has been trying for years to tell us all about itself, but only mothers had sense enough to listen. Even though mothers had the suffrage years ago with all women, people were not listening to the mothers much more than to their tiny tots. Even though we now know that the infant entered the world of the adults long ago, many sentient adults had not and mayhap still have not entered the world of the child.

When writing and editing *The Competent Infant* (Stone, Smith, & Murphy, 1973), the editors originally intended to cover the whole field of developmental psychology, but they twice reduced the extent of their coverage. After the last infant explosion, the editors found their book was fulfilled after only the first 15 months of babyhood had been covered. The day of the baby had come to stay, and the day of the baby is having its say.

The Wisconsin Primate Laboratory love research has added to the understanding of the significance of the first few years of life. The earliest three love systems, the infant, maternal, and paternal loves, function famously from the minute of birth, and their functioning fashions the form, the manner of faring, and even the fact of existence of the other love systems throughout life.

We have discovered the importance of the linkage of the five different love systems primarily through the dire and drastic effects of loss of any of the five through deprivation or privation. These will be discussed at length in the chapters on infant-mother separation and on the effects of social isolation. Experimental results can be best understood by considering separate behaviors which contribute successively to each or most of the love stages.

The bodily contact given by the mother to the infant or by the infant to the mother is such a normal and natural occurrence that we take it for granted. The reciprocal bodily contact between father and baby has not as long a record, but both father and baby are now putting us straight. The recent research of many investigators insists that the role of the father be given more recognition. In a study of first year behavior in the home, with both parents present with the progeny, Lamb (1976) found that whereas the mother held the infants more often for comfort and care, the father was more active in holding the infant during play. The holding behavior of each would reinforce the acceptance of bodily contact, but each in a different desirable manner.

If there is a severe lack or loss of contact comfort during these first loves, it will be extremely difficult for the baby to accept the next normal bodily contact when the time arrives for social play and peer love, the great socializer for the balance of the individual's lifetime.

The lack of early, pleasantly reinforced bodily contact acceptance would, at the same time, curtail agemate association formation by putting a damper on social play. A dearth of agemate friends and affectional ties would, in turn, influence the beginnings of adolescent friendship and sexual interests leading to long-term loves. Contactless sex is not worth the type or tape to type it.

The baby monkey has an enormous need and necessity to search and explore, even before the mother bonds are beginning to relax. Indeed, the first object searched and explored by the neonate is mother herself. The baby explores the maternal form with the mouth while nursing and, later, just before being rebuffed for biting. The infant explores with hands and with eyes (Fig. 70), with smiles and gurgling wiles. The eyes find other eyes, the smiles produce smiles coming back, and gurgles turn into sounds like mother makes. All of a sudden baby is communicating.

Children raised from early infancy in inadequate orphanages or foundling homes have, by countless researchers, been found to suffer learning handicaps of many types, in social relationships including play with peers and also in intellectual accomplishments. A study by Provence and Lipton (1962) definitely delineates the lack of communication existing in such institutional rearing. The attendants in the nursery are usually too few and too far between, too rushed and too pushed to stop and look and smile at the infants confined in their cribs, far too busy to coo and comfort. By the time

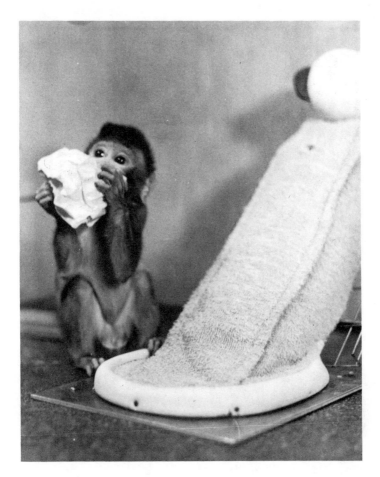

Fig. 70. Holding on to mother with one foot, the infant explores.

the baby has any real contact with other babies, sometimes as late as 18 to 24 months of age, the infant has a heavy handicap impossible to totally overcome. The only surcease would be a very early adoption, preferably before the age of 1 year. Extensive therapy is as unlikely as was the original availability of communication.

The importance of all types of communication is reflected in the effects of early maternal-child interaction. Mother love is not obtained by putting a quarter in a vending machine. Both mother love for the baby and baby love for the mother result from many variables promoting mutual reaction between the mother and the child. Likewise, the more opportunities which can be

Fig. 71. Eye contact communication while nursing.

discovered to promote interaction between the father and his offspring, the more apparent will become the father and baby loves for each other. One more conduit for all important communication will be found; the more likely will be the discovery that paternal love is strong and as much dependent on the underlying reciprocal variables waiting to be exploited as are both the maternal and the infant.

The basic variables producing maternal acceptance are the developing "feedback" mechanisms of the infants themselves. When these infants begin to smile at and with the mothers, they promote a mutual feeling of pleasantness and warmth. The normal development of maternal affection to her infant offers insight into the nature of the human autistic child. The autistic child is an infant toward whom the human mother has difficulty in developing affection in spite of the fact that this same mother has established normal maternal relationships with older or younger children. The infant cannot, for as yet not well known reasons, communicate or interact with the mother but goes its own autistic way, leaving the mother's affection no place to go.

When both incoming and outgoing messages from communication through gestures, eyes (Fig. 71), touch, and sounds are eliminated, communication ceases to exist, and loving, learning, and living are affected for an indefinite rest of the lifetime.

During the discussion of the variables underlying infant and maternal love, we repeatedly mentioned the importance of the development of basic security and trust which emerges from the aggregate of felicitous feelings created by the contact comfort, warmth, rocking, and composite contributions of the feeding situation. Because of the basic security the babies have the confidence to explore the world, knowing that mother will welcome them back to her arms if fear appears.

Fig. 72. Peer play and exploration.

The basic security leads to the social security through peer play which again broadens the base for later loves (Fig. 72). In addition, the ministrations of the maternal love system prepare the progeny for a different type of confidence, the confidence of handling its own body for leaping and climbing, and jumping and swinging.

Neonate monkeys climb up the mother's body as soon as they are born. A baby monkey has no inborn STOP signal and would keep right on going over mothers shoulders if left to his own urges and splurges. It would keep right on going over the top of a mesh wire incline as well. Fortunately, when climbing up mother's loving body there are mother's loving arms and libational charms to tell him where his luck lies.

The mother monkey will not allow the infant to try too much too soon, but when the time is ripe no mother monkey would ever dream of preventing the infant from going out to play (Fig. 73). The first play, even though other monkeys are around, is primarily individual play of jumping and leaping, running and swinging, in which the baby completes mother's lessons and learns a few of his own by way of what he dare do and what he doesn't.

It is this knowledge of learning what the body as a whole can do, in air and sea, on land and snow, that reinforces the first pleasure in mastery and also gives the first confidence in self, self without mother and self with approbation of peers, perhaps, but self all by itself. Peragration or self-motion play has proven to have a pleasure all its own (Mears, 1978). Equipped with this personal power and physical pride, the individual primate. human or subhuman, enters the socially wider world with a joyous jump.

The basis for self-esteem is a prerequisite for the love of living and loving no matter what love at what age or stage.

If there is to be any lack or loss of one of the loves, the timing of this lack tremendously affects the effects of the loss. For instance, peer experiences and peer love without mother love can produce three alternative anomalies, depending upon the time they occur. If peers are raised together from birth on without a mother or other live surrogate, the clinging to each other for contact comfort overwhelms the babies. They don't know when to stop clinging any more than they know when to stop climbing. They haven't even enough sense without mother to know when to start playing.

Our motherless mothers who themselves knew no mother love were a little better mothers if they had had some association with agemates somewhere along the line and, though it was better late than never, it was better early than late. However, without any mother love or any peer participation they proved to be beings completely devoid of loving leanings. Although normal monkey mothers are more than acceptable models for many human maternal ministrations, these motherless monkeys could be a model for only the most regrettable human mother of all, the child abusive maternal model.

Fig. 73. Mother, may I go out to play?

Since each of the five love systems depends, for the degree of its fulfillment and for its very possibility of fulfillment upon the full functioning of the antecedent love system or systems, the best solution of love's labors lost is not "Never to have loved at all" but to have, instead, loved them all.

CHAPTER II

THE BASIC TRYAD

The basic family tryad consists of father and mother and the baby, whom we allow to be of either one sex or the other. No matter how much we may have said about baby and mother, there always seems to be more to be said, and father has begun to come into his own position of importance in the scientific family as well as in the world erstwhile of men.

Mother love accomplishes many things in addition to the gifts of which we learned through the terry cloth mother and the neonate rhesus. Some of these accomplishments are not evident immediately after birth but must wait until the infant has reached the proper maturational stage, and the techniques which mother uses are sometimes clear and sometimes almost clandestine.

The primary thing that baby love or infant love for the mother accomplishes is tired and neurotic mothers. This can be known as the mother do-or-don't dilemma. More seriously, baby love is very similar to mother love in terms of its underlying variables. However, the things that infant love accomplishes or achieves are very different. They are more than just responses that differentiate baby love from mother love.

Infant attachment to the primate mother is both similar and different from infant imprinting in avian and probably ungulate forms. Infant-mother attachment (Fig. 74) in all forms appears shortly after birth, is extremely specific to the maternal object, and persists for a long period of time. In nonprimate forms, imprinting is a following response—ducks, geese, and probably all ungulates waddling or cavorting toward and after the maternal figure. Thus the infants make great strides toward the maternal figure, even though they do not always catch up.

The infant primate baby does not follow, but clings. The monkey can cling no matter what mother is doing or where she may be meandering. The human

Fig. 74. Reciprocal and enveloping affection.

infant can clutch, but mother's arms must do the cradling or holding. In birds and ungulates, the maternal attachment of following is the responsibility of the infants. The infant does most of the work, and there is no need for baby sitters. In primates, the mother does all the work entailed in spatial peregrinations. However, the human infant has its own special method of following, much sooner than most mothers may think. Within 4 days after birth, neonates are visually able to follow and track objects (Greenman, 1963), and a very few weeks after birth the baby's eyes follow mother when she leaves the room, a fun form of following for a beginning.

Fear affects the infant-maternal relationship in different fashions for the birds and the ungulates than for the primates. The appearance of fear of strange animate and inanimate objects implies a familiarity with other objects (Hinde, 1974). When objects in the maternal and infant world reach a certain degree of familiarity, the imprinting set begin to react with fear toward anything not among these familiar forms. When they fear they flee and when they flee they cannot follow even the waddling maternal figure.

Unlike the ungulates and the birds, primate babies rush to mother to be cuddled and comforted when afraid. If the human baby is too young to toddle, she cries a vocal plea for solace.

When we think of contact comfort, we usually think, and often refer to, what the contact does for the infant, but if the mother has been blessed from babyhood to marital bed with the enjoyment of cuddly comfort and a more adult bonus or two, the mother herself loves the very feeling of the bundle of baby.

The baby can do something for father, too. Fathers often have a feeling of being very much in the way after the birth of a baby, and fathers don't flourish by feeling that way. Fathers have recently scientifically confirmed the fact that they are not negative to neonates (Lamb, 1976); they themselves indeed so state the matter without evasion. They have positive affectional feelings toward the baby, a paternal bond, and see the tiny one as already an individual with a private personality whether perky or placid.

What the baby can do for father is to find the perfect paternal place and functions where he feels needed during the first few months of the baby's life. In an overview of background literature on father, the forgotten man, Lamb (1975) emphasizes the part played by the father, literally played in play with the child from 8 months on. As a matter of fact, it seems that mother may not interact with an infant anywhere near as much as most people do think. She is too busy with the feeding and care of the child, as I am sure many mothers have told us throughout the years.

The baby shows that play with father is appreciated, and probably is ready to be played with almost from the very beginning. It seems like a much nicer way to get started with life with father than having him change diapers, even if you can now get them without any safety pins which have negatively conditioned the fathers in the past. It was probably those sadistic safety pins that started the myth that fathers will leave the baby all to mother's tender loving care.

The neonatal and even infantile animal is helpless against any and all environmental threats. Without a mother, father, or other living surrogate, life would be short instead of sweet. Maternal protection responses accomplish more than the assurance of physical safety. The frightened infant rushes to the mother and, being wrapped in maternal charms, the fear if unreal is allayed, displaced, extinguished, deconditioned, or reconditioned. Part of this process, which has been going on for million of years, is now called behavioral therapy and is considered strictly confidential and restricted as part of the psychopathologist's armamentarium.

The mother doesn't realize how secret her techniques are supposed to be. She is in reality teaching her offspring of what it should be afraid, the dangers and strangers, and what it may accept as natural and normal though new to the babies. Novak (1974) found that real live mother-raised rhesus juvenile monkeys showed many fewer abject frozen responses to noisy nasty monsters

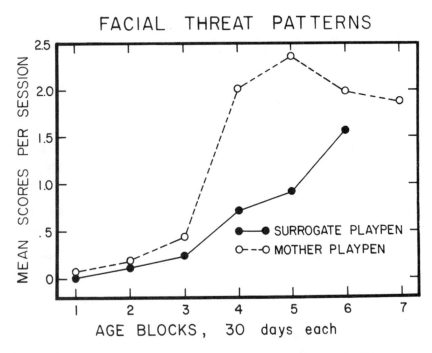

Fig. 75. The real live mother teaches threat gestures to ward off harm.

Fig. 76. Fear finds mother's arms.

with flailing arms and flashing lights than did the juvenile monkeys raised on surrogate mothers or with infant peers and without any mother. The mother-taught monkeys threatened back at the monster and when nothing happened just kept threatening automatically while curiously observing the contraption. Novak's study confirmed the findings from an earlier, less extensive study (see Fig. 75).

In learning about the linkage of loves, we found it is fortunate that the maternal-infant mutual love develops before the complex fears appear. A monkey or human mother fulfilling her functions can change fear from a socially destructive force into an enormously positive force for the infant.

Fear first drives the infant (Fig. 76) to the mother, into her arms or behind her skirts or jeans. The infant fastens and fixates on the mother while full of the feeling of security and keeps within easy reach during the period in which the mother can teach protective facts of fear. Whereas the infant's knowledge of protective behavior is founded on fear, it does not founder on fear.

Cloth surrogates can impart basic security and trust to rhesus babies but are socially ineffective when the infants reach the age of socialization. Terry cloth surrogates know no gestural language with which to communicate with the babies.

One of the most frequently used tools of facial gestural language of the rhesus mother is the fear threat face (see Fig. 77). During fear teaching

Fig. 77. The fear threat face.

Fig. 78. The silly grin.

sessions with the young, the monkey mother holds the baby tight against her body, where the baby snuggles safely. The mother, then, without any urging, puts on her fear threat face and turns it on full force at the approaching form, be it stranger or other danger.

One of the less frequent but most famous of mother's many faces is called "the silly grin." The gestural gyrations become more directive and definitive as the infant's insistence on escape and exploration explodes. Even after the baby monkey has escaped into the playpen play area into which the mother cannot go through the very little door, the macaque mother has retrieval ruses as last resorts. The silly grin (Fig. 78) is one. It works in over 50% of the

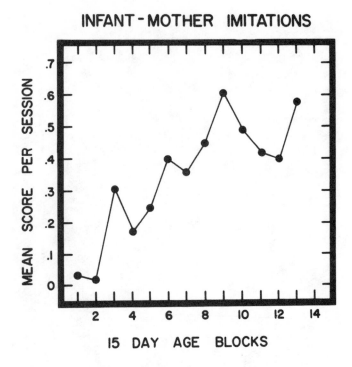

Fig. 79. Learning through imitation.

cases, and the little one comes in. In these instances, I imagine the infant is the one with the silly feeling.

The gestural language is vital for imitative learning of the baby macaque, whether the necessary knowledge concerns fanciful vs. factual fears or social success instead of social sorrows (see Fig. 79).

When human infants, for any reason, need to be entrusted for the major portion of their time and care to other than mother or father, it is wise to ensure that the caretaker not prove to be only a wire mother or even a cloth surrogate.

If human mothers are unaware of dispensing protections from terror, they need reminding of how many real protections they have taught and how many unreal they have comforted away. Every day some human mother looks out the window and sees her 6-year-old son engated in dire combat with the neighbor boy. Immediately, she knows that the grievous assault was initiated by the neighbor child and she rushes out to chastise the intruder or at least to protect her progeny.

It is entirely possible that early effective play is another social behavior facilitated by mothers, although recent research finds that the father may play a larger part in play than has been acknowledged (Lamb, 1976), at least when both father and mother are at home with the child.

We do know that children raised without adequate caretakers, mother or father, even with peers only, whether rhesus in cages or human children in some orphanages or foundling homes, show severe deficiencies in their play behaviors as well as much longer periods of time before acquiring basic play. Without adequate caretakers, the growing babies lose many prerequisites gained from love, whether love from the mother or father or other. They lack the contact comfort, the loving warmth which gives the sense of security that

Fig. 80. Damon and Pythias.

Fig. 81. The choo-choo effect.

allows exploration out into the strange world. They lack the early loving interactions of eyes and voice and lips and behavior itself which makes life worth living. In other words, they lack the urge to surge in any direction. They have ennui instead of joie de vie.

We call our Damon and Pythias monkeys (Fig. 80) the rhesus raised with agemates alone and without surrogate mothers on whom to cling, or real mothers to whom to cling and from whom to learn. With these baby rhesus, all play is handcuffed. Through their constant clinging they seal their chances for socialization. Larger groups left together too young and all alone exaggerate the together-together effect by creating a choo-choo (Fig. 81). Unfortunately, this choo-choo neither chugs nor conveys nor allows for play. Clinging love has cemented their love.

Many human working mothers have acquired a deep and determined appreciation of the infant's need for mother love. This need indeed does not extend to exclusive symbiotic smothering with love, and a little love will go a good judicious distance. Working mothers may find, however, that more manipulation of their available time is necessary to arrange and manage adequate early play supervision, as well as the play itself. If left with purely peer groups, unregulated, too often and too early, infants will invariably develop abnormal contact play or, more likely, contact nonplay patterns.

Finally, a basic maternal accomplishment is, strange as this may be, the encouragement of infant independence from the mother or, as we like to call it, unimprinting. In imprinting, as it was first described by Lorenz, the infant animal attaches or follows the mother, and this learned or imprinted response ties the infant unalterably and theoretically eternally to the mother.

Primary functions of the mother are those of guiding or guarding the infants, but guiding and guarding must not go on forever. The contact comfort bonds that tie the infant to the mother are in no sense an unmixed blessing. If these bonds are limited to infant-mother ties, they would, of necessity, produce a dark and dire maternal fixation, such as that found in stifling symbiotic relationships between the human mother and child. A single love may be enchanting but, if static, become antisocial, because society is composed of many and not of just one.

The independence from the mother is caused by joint efforts of the mother and the infant which conveniently coincide at approximately the same and proper age of the offspring. The infant becomes more and more interested in exploring the inanimate and then the animate world, and play begins to beckon. Mother encourages these initiatives unless danger lurks, and she at the same time discourages too frequent returns to maternal arms and charms. Even the male monkey babies learn that mother likes maturer men.

By maternal sanction of sorties of the infants into the outer social world, mother paves the way for the baby-love for the mother to generalize, as love should do in a society of others.

We know that early play predicated on a background of even earlier love proves to be the best breeding ground for developing desirable, positive social behaviors. There is a dire dearth of dedicated personnel to take over the vital maternal or paternal role in orphanages, or in day care centers and nursery schools primarily in existence for the use of children requiring full-time or supplementary care.

These personnel are needed not only to supervise playgrounds to prevent physical harm—that is another prevention, a more immediate, more discernible one. These are to prevent the development of antisocial behaviors, including and especially anticipating the development of aggression which develops later than the primary loves and later than the beginnings of play. Some tax money tagged for programs presumably but unrealistically designed to rehabilitate criminals might be diverted for such targets.

Of course, aside from the mother, no able or better baby sitter could be desired than the father, and many mothers and fathers are beginning to create dually directed and shared tryads, or foursomes, or more for some.

Fathers, not limited phylogenetically to Homo sapiens, have not been consistently conspicuous for the extent and the excellence of their baby tending. They have been somewhat more effective in caring for older infants but still quite chary in choosing their chores. Even among the Japanese macaques, where the fathers outshine their rhesus relatives, the stellar roles were limited to but a few of the many troops (Itani, 1959; Kummer, 1968) and cultural factors are given the credit.

Although paternal solicitude does not frequently include the holding, care, and comforting at which mothers excel, adult male rhesus do assume the protection of both infants and young, according to a consensus of opinion.

The creation of nuclear family accommodations on a permanent basis offered the opportunity for a longitudinal study of paternal behavior at the Wisconsin Primate Laboratory (Harlow, Harlow, Eisele, & Ruppenthal, in preparation). The nuclear family apparatus (Fig. 82) includes four large living cages and a central play area. Each living cage houses an adult male, female pair and their infants. Only the infants have access to the play area and the other family units, through child-size exit or access doors. The parents have contacts with the children of all four families.

This arrangement provided opportunities unlike any prior laboratory caging and was equally different from free-ranging feral living. It was, in many ways,

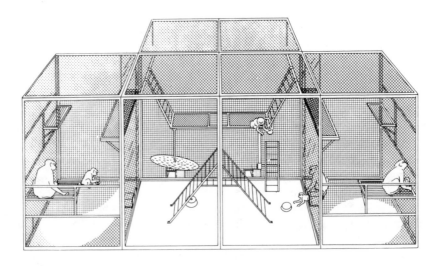

⌐————⌐ : 1 Foot

Fig. 82. The nuclear family apparatus.

Fig. 83. The nuclear family father.

a chance to observe father when he was without the pressure of his large-scale protective profession, or, in other words, father away from the office. He did not neglect his protective role. He still shook the bars of the cage and threatened infants in the playpen hostile to his own. He even persisted in this capacity as the infants all gained in status.

But father did not bely his dignity (Fig. 83) and take over from mother what was and had been mother's, the contact comfort, the restraint within the cage, retrieval from the playroom, and general care. In all 12 families studied, there was no question as to the dominance of the father. Without physical force, he enforced his control by the terror of his threats, by chasing, and by pseudo swats and nips. Infants might make faces at him, but only at a distance, and they did not defy.

ADULT MALE PLAY INITIATIONS

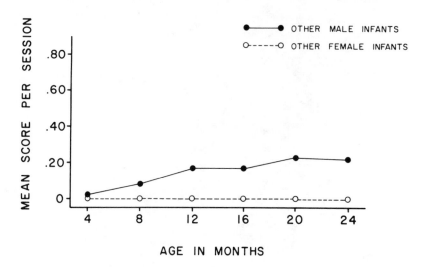

Fig. 84. Paternal initiation of play with offspring of other nuclear parents.

Play with the rhesus father presents a progressive picture with a steady increase of paternal initiation (Fig. 84). The adult males had a signal all their own, a dazed and glazed mesmerizing stare and a flimsy flip of the ear, and play was on. The fathers played with both sexes of their own offspring, but primarily with the males of other families. Little girls showed what might be called feminine inconsistency, although I am sure some better explanation will be found. When the little girl rhesus initiated play with fathers not their own, they renounced the wrestling and soft mouthing of play with their own fathers and, instead, they teased and pulled the hair of the adult males, the feminine equivalent of little boys' hitting.

Little boy rhesus like to play with any of the adult males, but the daughters follow their father's fancy and prefer to play with him.

Given the chance, the human baby proves its pleasure in the presence of father by making him a frequent choice when asked in gentle, guileful ways who loves whom. Of course the preference during the second year of life is primarily one of the man for the boys while the girls stay closer to mother's side (Lamb, 1976). The fathers interact more with the little boys at this age and these relationships may well be of importance in the development of sex roles in the young male. Boys separated from the male parent in infancy suffer deficits in both sex-role and maturational development (Biller, 1970).

The father may prove to be, for all time, the person most famous for his absence. In the complexity of human behavior, it is often very difficult to place responsibility for a certain behavioral development on one parent as opposed to the other. Contradictory evidence has been presented throughout the years. This investigatory double-talk confused not only the understanding of parental influence but also of the influence of culture and heredity. (Lamb, 1975) has presented thorough overviews of background information on parental effects on children's behaviors and has proceeded to separate, as far as possible, the influence of each parent according to the sex of the offspring. Except for specific studies of play behavior, the father's functions and his effect become much more evident by what happens to his sons when the father is in absentia. Not only do these sons lack in masculinity, but they are also less aware of moral and societal standards and are more aggressive.

If the father is home, he may be either a very bad father or a very good father, and is not as likely to be either of these extremes as to be just as inbetween father. If the father is always absent, you can be sure that the son is not going to have a good father on which to model himself, but if the father stays at home, you cannot be sure what the model is going to do. Fathers with a warm, loving relationship yet with strong masculinity themselves, fathers interested in babyhood from the beginning and with a nurturant attitude and personal concern over the birth period seem to offer the best alternative to the absent father in informing us as to the father's desirability.

Fathers are definitely designated as preferred playmates by human babies as young as 8 months of age, and on through the second year. With father absent, play between father and son is aborted. The way in which the father rhesus plays with his nuclear family children may offer a suggestion of the amelioration of aggression when father is home to excercise his play prerogative. The father monkey adjusted his play to a manner even less rough and rigorous than the respective physical states and age differences might require. In turn, the juvenile monkey lads played less vigorously than in play bouts with their own sex agemates, bouts which sometimes progressed to fears and hostilities.

When play with its socializing affect is aborted, aggression awaits the opportunity to take over.

To judge by the recent research and literature, the father has now caught up with history and with the mother and infant to assume his traditional role in science as well as in the family. Whether his role in the family remains traditionally dominant or delightfully different is the next question.

CHAPTER 12

PEER PERSUASIONS

The agemate or peer affectional system has already been designated the most important of all love systems. This is a brave but not a brash statement, since the peer system offers not only the widest range of affectional bonds but also the longest lasting of the lifetime. Peer loves start with agemates but may generalize to a broad range of different age levels, persons who are peers in interests or abilities or skills. The bonds may bind members of the same or opposite sexes and may outlive familial ties. Because of the wide range of age and wealth of personalities of both sexes which may be partners of peer bonds, it would be facetious to place rigid age limitations on the peer love system. There are, however, periods during which peer affection preempts the thoughts and actions of the developing individual. Parents are frequently and frantically dismayed when adolescents turn away from both the parents and parental precepts to choose instead the constant company of confreres, other teenage friends, members of a school club or adolescent gang. Parents cannot comprehend that teenage ties could suddenly acquire greater emotional significance than longtime love for the mother and father. To them the value systems of the darling daughter or stalwart son seem to have changed

189

Fig. 85. Early oral manipulation and visual exploration–and the human model.

overnight right before their eyes, or, more likely, right behind their backs. The growing youth are experiencing a back to nature movement as opposed to the hitherto effective parental nurture system.

Agemate loves do not develop overnight but are present in varying degrees once infants discover the existence of other infants. Eventually, given proper opportunity and nourishment during developmental years, they involve the widest social contacts and greatest social significance for the individual. In a culture where divorce is common, peer loves may outlast several heterosexual loves.

The agemate affectional system of the rhesus monkey begins to make its appearance when the intimate physical attachment bonds between the mother and infant first seem to weaken and waver and the mother permits her offspring to wander out of reach of the maternal arms. By this time, the infant has achieved a certain sense of security and the real monkey mother has also imparted to the neonate a rudimentary need for caution in the outer world. The first tentative moves toward infant-maternal separation are probably due just as much to the curiosity of the infant as to permission from the mother. For some time the monkey baby has been peering around the maternal body and the restraining arms of the mother, seeing as much as it can of what lies beyond. This same phenomenon is now accepted as customary human baby behavior, except that the human baby most frequently learns to look over the mother's shoulder. The shoulder becomes more than a breeder of burps as soon as the baby learns to hold his head up high. When fussing or crying, the human infant is known to quiet when mother holds her in such a fashion that she can see what is happening in the room behind mother's back. Let mother slide her down where her vision is obstructed and the fussing will usually return. The tilted carrying basket acquired immediate favor over the flat one from which the baby could not peer out, and the Indian papoose rode high, wide-eyed and content in his cradleboard.

The initiation of independence of the infant monkey from the mother is characterized by a meeting of minds. The situation is not a simple one for monkeys, and for human infants may be one of considerable complexity (Hinde, 1975).

As soon as locomotor capabilities reach sufficient maturity, within the first month of monkey life, the monkey actively and physically pursues his visually stimulated interest in objects away from mother. Since oral mechanisms of manipulation are functional at birth, oral contact and manipulation await only the finding of objects to manipulate (Fig. 85). At first the infant macaque may rush back to mother for intermittent reassurance until he discovers that the object is safe and he himself sound. It is doubtful that the rhesus monkey ever attains the digital dexterity of the human being, but his hand and finger gain prehension capabilities much faster and his grasp reflex occurs voluntarily by 15 days. Crude manipulation of objects is operant by 51 days of age.

In the course of finding inanimate objects to manipulate in a feral environment, the infant monkey naturally encounters other animate objects, and the same can be arranged in the laboratory. If congenial and of the right size, these new-found living beings receive the same oral and manual exploration and curiosity as the primary precursors of social play. As exploration encompasses utilization of objects, so also does social interaction become more complex. Recently, however, in an experiment conducted with Mears (Chaper 9), a clear-cut precursor of social play was found to be self-motion play, which is primarily an individual play pattern but a pattern of power which persists throughout every age and stage of life. Like all new behaviors, including social play, self-motion play receives its initial impetus from curiosity and an exploration of the new and strange. Once initiated, however, peragration provides its own reinforcement through the enjoyment of the activities. The pleasure of swinging, leaping, running, and jumping is the pleasure felt from the motion of the body through space. As the human child grows, this joy in peragration carries over to skating, bicycling, dancing, and skiing, all play behaviors which may be individually as well as socially appreciated. Social play first develops later and more gradually but in conjunction with self-motion play and stimulated by the apparatus used in self-motion play.

A still more recent, as yet unpublished experiment, by Mears et al. further elucidates and clarifies the relationship between self-motion play and interest in others of the same species. Because the development of self-motion play as an individual behavior is so dramatic and significantly rapid, we tested the infant monkeys of the original group to see what would happen in the familiar playroom with all of its beautiful equipment if each monkey infant was placed in the playroom alone for an hour on certain days of each week. Whereas in this new experiment play continues at a comparable pace and variety when the infants are in a group, play, with and without apparatus, plummets to almost nothing when the rhesus is alone. Whether or not the infant plays with the other monkeys when they are all in the playroom, it is very emphatic about not playing without the company of other monkeys. Interaction socially is not at all a prerequisite or component of awareness of other social beings, but the presence of other live monkey beings is already a necessity to insure a happy, well-adjusted young monkey. From the screeching and protesting, an observer would have concluded that the infant could not, as well as would not, play alone.

Whereas play, then, is the basic mechanism in the development of the peer affectional system, peer love, itself, begins even before social play matures. Exploration and curiosity play a large part, just as they do in the instigation

Fig. 86. Rough-and-tumble play.

of any new behavior or series of behaviors, but other basic prerequisites emerge and clarify the picture of the development of the complexities of social play.

While relatively gentle social exploration, such as touching, tentative tail-tweaking, and casual contacts are becoming familiar and self-motion play is spiraling, patterns of more active social play are beginning to develop. Our data clearly indicate at least two basic social play patterns, which we named rough-and-tumble or contact play and approach-withdrawal, noncontact or try-and-catch-me play.

The first of these play forms, rough-and-tumble play, is the most easily described and measured. Rough-and-tumble includes wrestling, rolling, pulling, and sham biting (Fig. 86). The intensity, tempo, and length of the sessions increase, but in spite of the fever and furor of the bouts there is no indication of pain or injury to the participants. Abnormal rearing conditions may distort the form and disturb the appearance of these responses.

Fig. 87. An invitation to rough-and-tumble.

A single monkey will initiate rough-and-tumble play by pulling another monkey's hair or gently biting at or on him (Fig. 87). Two monkeys are usually involved and the frequency of reciprocity of the play behaviors increases with two or more infants participating. Other variables affect both the frequency and the intensity. The presence of a cloth surrogate who imposes no restrictions upon the infants in any way and play sessions short in length both seem to be conducive to maximum rough-and-tumble play. When apparatus in the playroom enhances self-motion play, rough-and-tumble proceeds at a slower pace.

We compared playpen behavior of two groups of monkey infants, one group of four raised by real monkey mothers and a matched group raised by cloth surrogate mothers. The real mothers restrained interaction with other infants during the first month of life and we imposed experimental procedures which accomplished the same purpose with the infants raised with surrogate mothers. Even so, rough-and-tumble play appeared with both groups during the second month and maintained a higher level during the third month.

A wealth of accumulated evidence delineates sex differences in frequency of engaging in rough-and-tumble play during the first 6 months of life. During this time rough-and-tumble is not an aggressive form of play, but rather a very lively activity with a variety of behaviors involving mutual contact. As it involves sham fighting, it is natural that sham or social threat would be a component. Females, during the first 6 months, show an extremely low incidence of threat responses; the converse is true of males. Hansen plotted the frequency of this first social play form as exhibited by the two sexes; at no time did the feminine participation equal that of the male, and the difference in frequency is statistically significant.

The second form of social play was given the name of approach-withdrawal play, although with the human child the name of tag is the most familiar. During early human childhood, other names for try-to-catch-me games include try-to-find-me variations such as run-sheep-run, hide-and-seek, or kick-the-can. With both human and monkey infants the play pattern is characterized by little bodily contact, rapid chasing of each other with frequent reversals of leader and followers (see Fig. 88). Often the catching does not seem to be as important as the act of following. A pair of monkeys may choose a particular platform and make a series of jumps, one after the other, to another platform or swinging bar, then back up the ladder again. The patterns of approach-withdrawal play are found more frequently than the rough-and-tumble in conjunction with apparatus and self-motion play.

The tempo of approach-withdrawal play allows for variations and intermissions. With the female's preference for a larger proportion of passive behaviors, the game of chase may assume a teasing nature during which she pauses in a come-hither fashion, then dashes away as another monkey takes up the pursuit. There may be breaks during which each monkey goes its own way, one swinging from a parallel bar and the other on a rotating wheel.

In both forms of play an observer can see the constant opportunities for the development of social roles and can actually observe the emergence of these roles. The leader one day may be the follower the next; but as the months pass, the roles of dominance and submission are apt to be held consistently by the same animal for progressively longer periods of time. Social taboos, informal but unwavering, are passed on from one player to the next, and the participants discover the social behavior necessary to be accepted as a welcome member of the group. The playful limits of wrestling, pushing, tugging, and sham fighting become practically separated from painful aggression. The human child, if allowed to enjoy free physical play, passes through the same periods of social learning. Self-motion play leads into social play, and later from free social play actively into self-motion play games with rules, from the toddler London Bridge and drop-the-handkerchief to the outdoor games of hiding and chase, to the formalized self-motion games of

Fig. 88. A pair of rhesus infants, one after the other for the moment.

tennis, baseball, basketball, and football. Each, in turn, introduces the player to more socially structured regulations.

From the fourth to the eighth month, the monkey's life is dominated by play, and sexual as well as social roles have their playful beginnings. The infant-mother love bonds were cemented by the infant's enjoyment of contact comfort from the mother. This pleasure in bodily contact leads to the acceptance of the many varieties and vagaries of bodily contact in the free, physical play of self-motion play. In turn, the combination of prior experiences prepares the individual for the natural and normal acceptance and ultimate pleasure in the physical contacts of sexual behaviors. Sexual posturing begins to occur as early as 4 months of age. Both male and female precursors of adult behaviors, childish mounts, thrusts, and presenting have their genesis in play. These increase in both enthusiasm and proficiency as the infants play and grow toward maturity. Initially, both sexes display mounts and thrusts, adult male sexual behaviors, and, likewise, both sexes exhibit the adult female sexual "present." When young, neither sex expresses these behaviors appropriately. At first the incipient sex patterns are infantile caricatures of adult patterns, but, as the young monkeys develop, the sex behavior resembles ever more closely mature sex postures. Gradually the inappropriate responses drop out and by the time sexual maturation occurs, between the ages of 3 and 5 years, the behaviors are directed toward the biologically proper recipient and in the anatomically proper direction.

Development of play among the laboratory monkeys raised by real mothers and with peers for playmates follows in the chronological footsteps of play observed among feral monkeys. This fact alone would not indicate the importance of play in the social and sexual development of the growing rhesus, but we have learned the real significance of play through the sad stories of monkeys raised without real mothers and without agemate play experience.

When monkeys are raised with agemates alone, without the mothering experience, they have no mother to encourage them to go out and play with their confreres. They have no mother to give them the sense of security to seek new adventures, to go out and explore and, incidentally, to find friends. Instead, they spend their time clinging insecurely to each other and they never do learn to play naturally and normally. By the time they learn to play at all, they learn too little and too late.

When, on the other hand, monkeys are raised with mothers alone and denied the opportunity to play, they become withdrawn into themselves, do not develop into normal social animals, and also are unusually aggressive. Needless to say, the monkey mother did not make this choice. The opportunity to play did not present itself and there were no peers available to guide them through play into the proper and pleasant social relationships.

Toward the end of the first year and the beginning of the second year of life, the rough-and-tumble play of monkeys becomes even rougher. As the play becomes tougher and more aggressive, it is more than ever the play choice of the male monkey who, then, obviously plays more and more with other males. The females, left more to their own devices, enjoy feminine company more often and begin to reflect the separation of the sexual roles. This is what Freud explains, quite differently, as his latency period. If the patterns of play are adequately understood, no repressions of oedipal incestuous and aggressive wishes need be imagined, no resultant identification with the parent of the same sex need result. It is this identification which, according to Freudian theory, leads to close association with members of the same sex when the human child is between 6 and 11 years of age.

Although the play behavior becomes more aggressive and there may arise some protesting cries from the players, no real injuries, no blood baths, or even had blood result. The fun and fury sometimes fuse, but the socially aggressive play never threatens to disrupt or terminate the agemate affection.

There is no monkey pattern of precise, unchanging, stereotyped form of behavior such as characterizes the aggression of the stickleback. The behavior of the monkey is much more complex, more similar to that of the human primate.

Because of the late maturation of aggression and the very early formation of peer love bonds, the friendly force of the peer group exerts a form of social control during play. There is intermittent shifting of dominant and submissive roles which ameliorates any expression of harmful violent aggression. The dominant individuals cannot maintain status without the support of group members, and the rhesus accept these relationships without recourse to lethal combat. The bonds of affection already present preclude expressions of violent anger toward the same individuals, and members of the group maintain mutual feelings of affection over long periods of time.

One other affectional system is at work in maintaining moderation in aggressive behavior. In the extended monkey family, which includes the baboon, one of the famous functions of father is the prevention of abuse of the privileges accorded to size, age, and the autocratic authority of relationship. When play becomes too noisy and disruptive, father steps in to protect the weaker, smaller, more submissive participants from harm. Subhuman primate fathers seldom frequent the golf course or the 5 o'clock bar, and tend to be available for crucial crises.

A group of rhesus monkeys approximately 2½ to 3½ years old had played together and acquired strong agemate affectional bonds, as monkeys who played together on the Madison Zoo monkey island should. The monkeys enjoyed swimming in the moat surrounding the island. There was a castle on the island and the monkeys would dive into the water, swim, and then return to revel and frolic on the rocks in front of the castle. They lived in a veritable paradise, but the zoo director decided to place a half dozen crocodiles 6 to 10 feet long in the moat. The infant monkeys would swim, only to suddenly find that they faced a yard-long crocodile yawn. Gradually they retreated, to spend 24 hours a day on the castle expressing to each other their opinion of crocodiles.

The monkeys tired of their banishment from paradise and they developed a scheme within the limits of safety and sanctity imparted by earlier maternal ministrations. A careful coterie of infants would form a line on the extreme edge of the moat and await the advent of a concrete-minded crocodile. As a crocodile neared the edge, one of the waiting monkeys would grab its near forelimb, a pair would seize the hind limbs and immobilize him against the rough surface of the moat's wall. Other members of the monkey menage would grab or bite the exposed side or belly. This may sound ghastly and gruesome, but was in reality causing no damage to the tough crocodile limbs. The main interest of these observations is that psychologists have experienced

Fig. 89. Group cooperation, albeit nonaltruistic.

great difficulty in producing cooperative behavior in monkeys and also in the more socially cooperative chimpanzees (Fig. 89). Conventional learning techniques had enjoyed limited success whereas our accidental success was undoubtedly bound to ingroup affectional feelings. Not only was the aggression cooperative, but also ameliorated and more mischievous than Machiavellian.

With the human child, you can understand how gang violence could develop if aggression is learned before the ameliorating effects of love have a chance to operate. Let there be a caretaker or a mother or surrogate without the time or the inclination to provide the basic ingredients of maternal love. Substitute daily TV with its inescapable violence for positive peer play. First introduce the child to a group of peers who have also been raised on violence and the power of peers remains without the persuasions of love.

The rewarded violence and exciting violent aggression of TV is now known, without doubt, to increase the likelihood of violence in the viewer (Berkowitz, 1971, 1973). No longer do scientists concur in the theory that expressed violence acts as a cathartic to relieve an individual of his violent feelings. Instead, the active violence of the youth upon joining an adolescent gang would be increased by peer approval of the teenage gang members and unrelieved and unadulterated by previous, positive love systems. To say that these suggestions are concrete and logical does not simultaneously supply a pat and simple solution.

It is obvious that agemate or peer affection among monkeys is not limited to the period of infancy, but continues to operate as a variable of importance in interpersonal relationships in the heterosexual affectional system. It is possible that the breadth of the heterosexual relationship within, over, and beyond sexual attraction and the lasting qualities of the relationship between human beings may depend upon the nature of the background of peer affectional ties.

CHAPTER 13

HETEROSEXUAL LOVE

The history of the psychological study of behavior during the closing decades of the 19th century and the first half of the 20th century could be plotted on the basis of the relative importance of heredity or environment. First the scientists enumerated and detailed the intricate inherited instincts forming the base of behavior, and then they turned to explaining how learning was responsible for all of man's behavior except for the simplest of unlearned reflexes. Educators staggered under the learning load. In certain cities and in certain states, the weight of responsibility for the learning load was shifted to the pupils themselves, young as they were. In primary education, great selective adeptness was credited to the child's ability to make choices of what to learn. The child often chose not to learn reading, writing, and arithmetic. Eventually the learning load doubled for these educators.

Now many people believe that culture has cultivated many behavioral differences between men and women. They believe that learning is responsible for most of the behavioral differences between men and women. When mothers who have raised both baby boys and baby girls insist that they are by nature diverse, the answer is that they are different because older people, mothers and fathers and aunts and grandmothers, have treated them differently from the very beginning. Today some mothers and fathers are conducting experiments to see if they can change little boys by having them play with dolls. The results will be of interest, but there is one result for certain, and that is that no matter the effect on the little boy playing with the dolls, this will change the attitude of all the little boys who still are playing cowboys and with trains and mechanical toys toward the little boy raised playing with dolls.

I have acquired the firm belief that simple innate responses produce simple learned responses, and that complex unlearned responses are at the root of all complex learned behaviors (Harlow & Mears, 1978).

Love between the sexes falls into this latter category, just as do play and aggression. Heterosexual love between sexually mature females and males is based upon a wealth of unlearned behavior, some of which depends upon successive maturational stages just as does the developmental change in height or the qualitative as well as quantitative shift in weight with age.

My belief is equally firm that there are very complex behavioral differences between men and women and that, on the whole, women will continue to prefer to love men and maybe marry men and that, on the whole, men will continue to love and marry women.

There are exceptions. Many people consider these exceptional people abnormal, just as many people used to consider geniuses to be abnormal. There is still no valid reason to consider geniuses abnormal and perfectly good reason to consider geniuses exceptional.

Of course, in using the term "heterosexual love," we automatically limit the love to love between the sexes, love fully fashioned after sexual maturation but developmentally influenced by previous maternal, paternal, and agemate affection which may exist through and even beyond heterosexual love.

Anatomical differences between the sexes are generally accepted by the majority of human beings, with the exception of those who wish to improve upon Mother Nature, God the Father, or evolution. Or those who might feel that the grass skirts are greener on the other side of the courts. Had God created man or monkey, male and female, with identical anatomical arrangements, there would never have been any need for historians and we would not be here asking questions.

There is a fundamental rule in biology about anatomical and behavioral differences, both within and between species: If anatomical differences do exist, far greater and more profound behavioral differences must exist.

In thinking about heterosexual love or love between the sexes, it seems less relevant to think about the differences either anatomically or behaviorally between the female sex and the male sex than to think about the great diversity between the members of the same sex, either to begin with or to end the discussion. I believe that love or affection between members of the opposite sexes is determined much more by the differences between members of the opposite sex than between men and women.

The term sexmates may seem to introduce opposite sexed pairs but if agemates mean mates of similar age, then sexmates must mean pairs or more of similar sex. Whether or not a man chooses the company of a certain woman because of her anatomical differences or because of her behavioral differences from her sexmates is more pertinent to the matter at hand or on foot, to keep on an antomical basis for the moment.

Of course, you cannot always be sure whether or not anatomical or behavioral characteristics are at the base of affection. As any human being addicted to puppies, kittens, cats, or dogs can tell you, real affection is not only possible, but extremely natural, between members of distinctly unlike species. Do you like the kitten or the puppy because it is soft, handsome, friendly, cuddly, or pretty?

When we started our careers in psychology, the animal laboratory at the University of Wisconsin had just been torn down, and I accepted an invitation to work with the animals at the Vilas Park Zoo in Madison. To anyone acclimated to a scientific laboratory for years of graduate work, the zoo seemed far from the scrubbed, strictly regulated, even though not sterile, sanctuary. But the Director did everything within his power to help, and most of the monkeys were fairly cooperative as well.

Except for Tommy.

Tommy was one of the biggest baboons I have ever seen. He was a Sphinx baboon, and he must have weighed 90 pounds. We were running delayed reaction tests. We would place two cups on the table, show the monkey a piece of food, and then cover it with one of the cups. The length of time the monkey could remember under which cup the food had been placed was measured.

Tommy was fine until the length of time of delay exceeded 10 seconds. Then he would hurl the cups away or grab hold of the table and crush it against the bars of the cage. We thought we would have to abandon testing Tommy.

Until Tommy met Betty.

Betty was a beautiful girl. She may not have weighed much more than Tommy, but otherwise her anatomical characteristics were just about as opposite from Tommy's as could possibly be.

Betty had come to the zoo to help us and she started to feed Tommy from time to time. He would reach out his arms and she would rub his wrists and hands. She let him play with the ring she wore. It was as simple as that. Tommy fell head over heels in love. Tommy had good taste.

Betty began to test Tommy at delayed reaction tests. There were long delays. During the delays Tommy would shake the bars and beat the floor of the cage. Then Tommy would look at Betty and his heart would melt. He never again damaged the apparatus as long as Betty tested.

To return to intraspecies heterosexual love, Beach has made in-depth studies of a voluminous variety of species of animals from flying fish to cantering colts. Now, when you think of breeding animals, you may think of breeding horses to emphasize the swiftest of the fleet, the best matched pair for the carriage fancier, or the smoothest jumper to take the hurdles of the hunt. With dogs it would be the best of the breed. The choices are usually man-made. Beach, the species specialist, was fascinated to find that some male dogs actually formed greater affectional bonds for some female dogs than for other female dogs, and some of the female dogs actually formed firmer affections and a preference to play and to produce with some male dogs rather than with other male dogs.

Psychology has always been the study of behavior, and anatomical singularities are of interest primarily as having a bearing upon behavior. Without any statistical analysis to order our opinion, we would say that the doggies' behavior had probably influenced the other doggies, and Betty's behavior had probably influenced Tommy, much more than any anatomical attributes. And Tommy was a great big baboon.

Many preference studies of human beings have been started, but they often bog down and become belabored because the complexity of human behavioral characteristics confuse and compound the results.

Developmental psychology has proliferated productively during the 20th century, and this is especially true of the last two or three decades. Many of the studies of baby behavior have contributed clues for the understanding of behavior far beyond babyhood. There have been studies of babies' preferences for many behaviors and there have been studies of babies' preferences for people or of subhuman primate infants' preferences for other subhuman primates. In both the monkey and the people studies, father has received more baby votes than might have been surmised, not as necessarily being more preferred than mother, but under certain conditions mother and father were each preferred more than the other (Fig. 90). Mother has been so loaded with love and lauded with love by poets and artists and novelists throughout the ages that father, as an object of special affection, has slipped into the shadows. For instance, when you mention owing a debt to your father, you are more likely referring to the gold of the realm rather than to his heart of gold.

One of the most unique and interesting of the baby studies (Lee, 1973) was the study of the social strategies and agemate preferences among babies 6 to 9 months old, all attending, with some parental cooperation, a Cornell University day nursery.

Fig. 90. Under certain conditions, the father is preferred.

Of this sociable group, Jenny, age 9 months, proved to be the most preferred and Patrick, 8 months, the least. The results are such that it is hard to believe that the babies were not yet gifted with speech, but, fortunately, there could be no introspection on the part of the participants to cloud the conclusions. Since Jenny and Patrick were of opposite sexes and one the most liked, the other least liked, we shall have to settle for people preferences, in general.

The most striking differences were found in the initiating and the terminating of contacts between our two finalists and the rest of the babies. Patrick was obviously the most aggressive in taking the initiative, in both the frequency and the manner of the initiation. He not only initiated significantly more contacts than did Jenny, but he did it the hard way, by grabbing the toys of other infants or by physically making contact.

You can almost hear the parents saying, "Isn't Jenny sweet?" Although more reluctant to start the social exchange, when she did, Jenny did it in a truly tactful, gentle, or as so many might say, the feminine way. She used passive looking, facial and vocal gestures, and more social signalling in both the initiating and responding situations.

Patrick may have been a go-getter as a starter, but he surely did not know when to stop, especially when he was the initiator of the behavior. Maybe it is just as well that language was not at a further stage of development. Jenny, on the other hand, usually was the one to terminate her own contacts, seeming to have sure knowledge of when her partner had had enough.

Had Jenny been the toy jerker and Patrick the pleasant politico, social sanction would no doubt have been reversed.

Whatever disparity in behavior does exist between the sexes, the overlapping in characteristics is also significant. Just as there are many women in the United States over 6 feet tall, inches above the average height for United States men, there are many American men shorter than the average height of American women. With wise dietary decisions and a further increase in the population of California, these average heights shall probably increase still further, but the overlapping will always be there. Behavioral characteristics, of course, follow the same rules as do the anatomical traits.

Having now mitigated the matter of differences between the sexes, we shall examine some statistical differences between female and male subhuman but similar simian relatives of Homo sapiens. (There are present preferences for the use of the term "person" instead of man, but that might convert Homo sapiens into persona non grata).

To start with a trait which might seem too complex to be found in even the varied behaviors of the rhesus monkey, do girl monkeys and boy monkeys differ in their responses to monkey babies?

To test this hypothesis, the response made by eight preadolescent male monkeys and eight preadolescent female monkeys were compared. The male and female monkeys had been raised in a limited social environment without any prior experience with any monkey infants. They brought into the test situation only the gifts bestowed by God.

The results were clear beyond question and dramatic in their decisiveness (Fig. 91). The responses by the males to helpless infants a half year or less of age were ninety percent indifferent or mildly rejecting and less than ten percent affective. To the naive little boy monkeys babies were for the birds. Since the males were prepuberty males, the babies did not even signify the birds and the bees.

Contrariwise, the prepubescent female monkeys, totally devoid of

Fig. 91. Helpless monkey baby.

experience   related to maternal ministrations, showed a strong, positive responsiveness to the miniature macaque infants. Almost 90% of their behavior with the infants was positive. They held, cuddled, and even groomed the little ones.

As we have already mentioned, human fathers have recently committed themselves in scientific symbols to feeling an affection for babies and an interest in being with them. The fathers, to date, have probably not been given an even chance to put this tenderness to the test of time and turpitude. However, we have yet to see reversed the relative frequency of female to male baby sitters or the relative frequency of brawny boys to lithsome lassies hard at work in the yard, be it brick, lumber, or lawn.

SURROGATE PLAYPEN
ROUGH AND TUMBLE PLAY

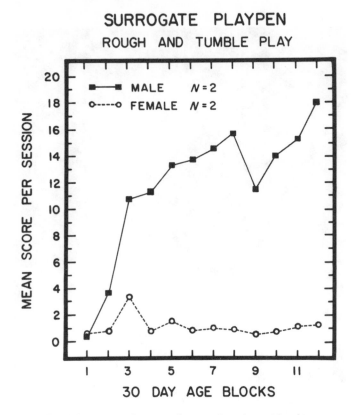

Fig. 92. Sex preferences for rough-and-tumble play.

Even the most critical culture adherents accept the fact that human baby boys are both heavier and longer than baby girls (Tanner, 1970), and that they are more vigorous in their bodily activity (Korner, 1974). These traits have passed the most exacting of the rigid rules and requirements since they are evident during the first days of life, before experience or culture can contaminate.

These significant facts are consistent with the Wisconsin research results on monkey play. After self-motion play has paved the way, two social play patterns gradually gain distinction, approach-withdrawal and rough-and-tumble sequences. Approach-withdrawal resembles human childrens' tag or chase, with oft-reversing roles of the chaser and the chased.

Rough-and-tumble play, with its wrestling, rolling, and noninjurious biting is preferred by the more vigorous male monkeys, and the significant preferences continue through testing from the age of 3 months to the end of testing at 13 months. The boy monkey not only engages in rough-and-tumble more often than the girl monkey (see Fig. 92), but also initiates this play form more frequently.

Two qualities found to be statistically more characteristic of the behavior of the infant female monkey have been given names which do not flatter or too accurately describe the actual activity. The first of these is termed *passivity* (Fig. 93), which overdoes the inference of submissiveness. The more meticulous description would be quiet, relaxed postures, behaviors interspersed with periods of livelier living.

The second, *rigidity* (Fig. 94), is a postural response made at the approach of another animal, with the body motionless, limbs rigid, and the head averted. It has no resemblance to its rhyming term, frigidity, so far seen only in monkeys raised from birth in life devoid of all love bonds.

Infant sexual posturing manifests obvious differences between the baby male and the baby female, although neither for a long time bear much

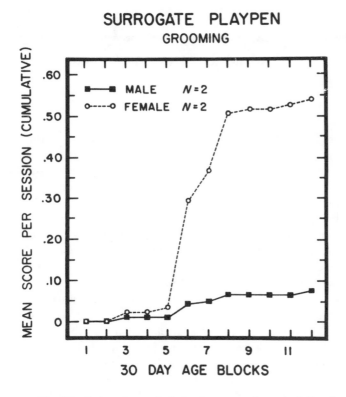

Fig. 93. Relaxed, gentle behavior more characteristic of female.

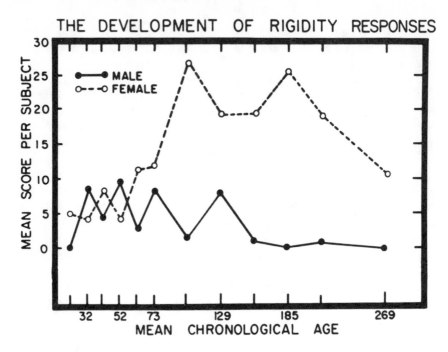

Fig. 94. Feminine postural response of rigidity upon approach of other rhesus.

resemblance to adult sexual behavior. Sexual posturing passes through developmental changes from infancy to sexual maturity, at which time consummation is first achieved. Likeness to the adult sexual behavior becomes more apparent as maturity approaches and tentative infant sexual behavior bumbles itself away.

The terry cloth surrogate mother has some very good qualities but she is not noted for passing on cultural, acquired information to her young charges. The diversity of the behaviors between the sexes is the more impressive in that they are present with surrogate-raised infants whose mothers could not possibly have taught the behaviors to them.

The threat response, in addition to being useful in fear situations, as taught to the babies by the real monkey mother, is a response also used in establishing dominance. Dominance and submission are two of the social roles duly developed in the playroom situations. Like rough-and-tumble play, threat has been found to be a predominantly male characteristic (Fig. 95). Not only is the behavior more frequent with the male, but there is quite a difference in the targets for the threats of the two sexes. Whereas the monkey boy babies threaten both males and females, they threaten the males more often. Female

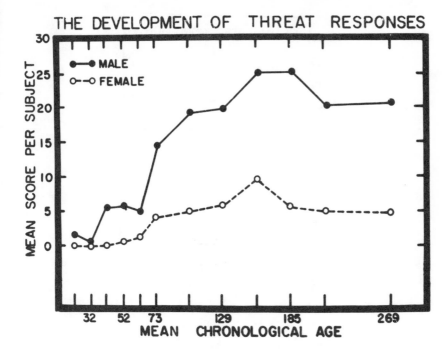

Fig. 95. Threat responses more predominant and earlier apparent in male monkey.

infants are even more selective. They may threaten other females, but they hardly ever threaten males. Unfortunately, due to a lack of the proper structures, monkeys are incapable of speech, and we shall never know whether or not the use of spoken language might reverse the threat ratio and the direction of threat.

It has long been accepted that human females, both small and large, are more precocious physically, physiologically, and behaviorally than boys. The differences begin prior to birth and continue well into adulthood. Girls smile sooner and more. On the average, they sit, stand, crawl, and walk before boys. They grow faster, and they mature at an earlier age. We know for certain that they talk sooner, although it is not as certain that they always have the last word.

The accelerated growth in female monkeys follows the same general pattern as in human beings. The only trait which becomes apparent earlier in the male monkey than in the female is that of aggression. Aggression rears its ugly fist at about the age of 2 in the male monkey and does not begin to appear until 3 or 4 in the female. Self-aggression (Fig. 96) is an even later maturing trait, but again it matures earlier in the male.

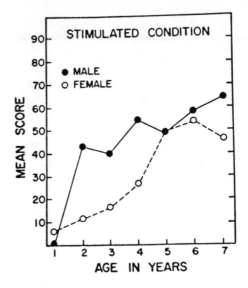

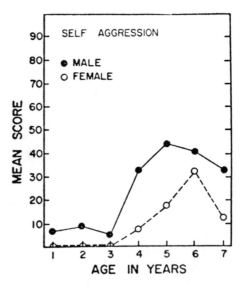

Fig. 96. Sex differences in the appearance and degree of aggression: externally directed aggression (top) and self-directed aggression (bottom).

Among human educators there has been such strong feeling that aggression is a learned behavior that, until many recent experiments, very little has been done to test the hypothesis. The late appearance of the trait, with the opportunity for learning, has encouraged the assumption of its being acquired and has at the same time made checking and corroboration more difficult. The importance or possibility of maturation of behavioral traits seems to be difficult to digest unless the behavioral trait is obviously anchored to anatomical or physiological maturation. However, all of the simian studies indicate that early love and early play can ameliorate the appearance and violence of aggression and that the late appearance of aggression is for this reason a most fortunate fact.

Not only is the early experience of maternal love, peer love, and peer play vital in the curtailment of aggression, but it is vital in preparing the individual, human or monkey, for heterosexual love. We had a two-fold purpose in introducing the subject of people preferences in general, rather than first discussing sexual behaviors per se. Freudians attach sexual significance to every affectional relationship without exception, be it infant-maternal relationships or later affection between parents and offspring. During the last half of the human or monkey agemate affectional system, boys and girls usually form and fall into dichotomies. Boys primarily seek the company of boys and girls primarily seek the company of girls. Freud called this the latency period, for which he accounted with a fabulous and fallacious theory: The girls and boys seek same sex associates because of repression of infantile sexual attractions of the boys for the mother and girls for the father.

The actual and factual reason for the separation lies in differences in social interests, particularly in play interests. Boys like to play rough, and get more rugged as they grow. They prefer other boys as suitable play partners because girls withdraw gradually as the game becomes rougher and turn to the gentler games and playing house like mother.

Freudians to the contrary, we believe that all loves transcend solely sexual attractions, and heterosexual love is no exception. The importance of sex is not to be minimized, mutual sexual participation and appreciation, but in order for this to be love as well as sex, personal parental and peer love as well as agemate play need to be present prior to passion.

The search for successful sex per se may lead to frustration or even to satiation without the knowledge that love needs sustenance from other sources as well as sex. Heterosexual love, in order to survive and persist past sexual consummation, depends on earlier affectional ties between the same and the opposite sexes. The breadth and depth of these ties depends upon the breadth and depth of previous experience as well as the acquisition of the ability to communicate with others.

These experiences begin with the neonatal period.

PART IV

THE PRICE OF PATHOLOGY

CHAPTER 14

PATHOLOGICAL
PERSPECTIVES

Within the short temporal span of only 2,500 years, significant gains have been made in the diagnosis, classification, and treatment of varied mental disorders. No longer is hysteria a wandering of the wanton womb, as described by Aesculapius, but instead it is a conversion neurosis. What has transpired over the last 2,500 years is that the interpretation has changed. It is not the physical disorder which has produced the psychic suffering; it is the psychic suffering that has been converted into the physical disorder.

The Greeks also recognized depression, and from time to time they had good reason to be depressed. However, the Greeks had no patented pills to alleviate their suffering. Today we have a wealth of patented pills, and some of the best abbreviate the period of suffering, whether or not they alleviate the pangs of suffering. We also have electroconvulsive shock, but no physician would engage in a treatment so shocking. However, it is medically acceptable if the treatment is called "electroconvulsive therapy." Shock is sadistic, but therapy is therapeutic. Quoting Shakespeare, "A rose by any other name would smell as sweet." We also have various verbal therapies which are as powerful curative agents for the alleviation of depression as is spontaneous remission.

Retrospectively, the brief and bewildered pictures which we have just painted are the perspectives of the past. The time-honored techniques of research and rescue will, with good reason, be used throughout the foreseeable future, but there is also reason to believe that they will be supplemented—and to a considerable extent superseded—by perspectives apparent at the present but only faintly fathomed for the future.

Research in psychopathology has gone far, and it will continue to go farther. The major psychoses are so disparate and dour, so bedeviled and destructive, that research on any kind of subject is justified, and mortal men of anguish and agony have often been the subjects of choice.

In most medical fields, exploratory and even definitive research is conducted on subhuman animals. This is accepted and acceptable if proper care, caution, and control are conducted to prevent animal abuse and affliction beyond the discomfiture which the disease entails.

Platitudes rather than prudence have restricted the use of subhuman animals as subjects for psychopathological research. One of the problems in the use of nonhuman animals for psychopathological research results from the common assumption that there cannot be madness without mentality, and certainly not human madness without human mentality. Diagnosis of human madness in terms of the classifications convenient for humans is difficult to make for human morons, idiots, and imbeciles. This does not mean that human imbeciles and idiots do not have human psychoses. They suffer from communication problems similar to subhuman animals. Perhaps behavioral studies of subhuman animals will reveal what the psychopathologies of humans really are, and this alone would be progress beyond the programs of Aesculapius. Similarly, it should be there are few, if any, basic human mental abilities, including basic language (Premack, 1971; Gardner & Gardner, 1971), which are not found in somewhat simplified form in nonhuman primates. The limits of subhuman animals as subjects for human-oriented psychopathological research will be determined by time and tests, and not by wands or witless wonderers.

At the University of Wisconsin Primate Laboratory we have developed and are developing a family of behavioral techniques designed to produce psychopathology in monkeys, with the goal of gaining understanding and insight into human psychopathology, both its cause and cure.

For more than a decade we have conducted researches on the induction of human simulated pathologies in monkeys. We would undoubtedly have studied monkey pathology much earlier had we not read the psychiatric literature. One of America's greatest psychiatric theoreticians had written to the effect that human psychopathological syndromes could not be reproduced or even simulated in subhuman primates, thereby showing the extent of the influence of dogmatic and wide-sweeping predictions. Perhaps our greatest, most significant discovery is that human behavior generalizes to monkeys whether or not monkey behavior generalizes to human beings.

Our intellectual latency period, however, gave us one great advantage since we used it to study love in all its fabulous forms. The recognition of love is of enormous value in the study of psychopathology. Depression rarely leads to love, but love frequently leads to depression as we all know. The human pathological literature demonstrates that important facts can be uncovered from inadequate love theory, as in the case of Sigmund Freud. Freud's researches on sex are of enormous practical and theoretical importance, and they would have been of even more importance if he had known their phyletic origin and where they were derived. During a serious discussion of affectional systems, Harlow used the term "love," at which the psychiatrist present countered with the word "proximity." Harlow then shifted to the word "affection," with the psychiatrist again countering with "proximity." Harlow started to simmer, but relented when he realized that the closest the psychiatrist had probably ever come to love was proximity. Psychiatrists typically look upon normal behavior as a deviant from abnormal human behavior, while the experimental psychologist measures abnormality from a normality bias.

Our earlier studies in psychopathology are related to infant and mother separation and to both partial and total social isolation from birth on, for 3, 6, or 12 months and more. We have tested the effects of brief isolation periods and prolonged periods. We also have monkeys that were raised normally for 6 months or more and then subjected to total social isolation. We call these monkeys our late isolates. Partial social isolation includes animals raised in such a manner that they can see and hear other animals but cannot mingle or interact with them. In total social isolation an animal lacks all contacts, visual, auditory, tactile and experiential, with other animals. In total social isolation this would include not only mothers but isolation from any members of the monkey group, agemate playmates, siblings, and fathers. Although our early separation experiments concentrated on the effects of separation of the infant monkeys from its mother, later research studied the effect of separation between agemate infants raised together without mothers. Our monkey-human analogues for infantile anaclitic depression are precise, and the technique for its induction by separation of infants from their loved ones has been very successful. Finally, we are continuing our long-term program on the induction of a variety of forms of depression in monkeys of varying ages. We are interested in simulation of psychopathic states other than depression in simian subjects. There have been suggestions that we may inadvertently have produced some sociopaths among our experimental monkeys and we know that we have, equally without intent, produced some schizoid-like behavior in individual monkeys. We now intend to try and create intentional, controlled replication of results. Eventually we may get a glimpse of some simian sanity parameters and their possible implications for relief of oft-told miseries, so far eased but not relieved, for the human counterpart.

Infant and mother separation research occupies a unique position in its relation to isolation studies. Isolation may involve total social isolation of the infant from the mother from birth on, for varying periods of time, whereas separation research studies the effect of separating the infant from the mother after affectional bonds have already developed. Another common classification distinguishing between these two phenomena is that of privation and deprivation. The later experimentation on the separation of infants from their agemates encompasses similar distinctions.

Our entry into the field of psychopathology was intentional but sparked by the creation of an animal totally incapable of becoming abnormal, the terry cloth surrogate mother. This was the same mother who had played the major part in displacing the Freudian concept of mother love, the cupboard theory that hunger and its association with the bounties of the breast constituted the primary variable underlying the love of the infant for the mother. We have known ever since that, although the baby hungers for love, it does not love from hunger. During the experiments establishing the actual variables on which mother love depends, we discovered that the maternal contributions to establishing within the infant a feeling of security and thrust could come from the presence of the surrogate as well as from the real mother. Indeed, the infant showed real infant love for the maternal substitute (Fig. 97).

Fortified by the knowledge of the powers of our mechanical mother, we tried to create pathological disorders in the infant by changing our cozy mother into a cruel caricature of herself. Rosenblum and Harlow (1963) had earlier found that dire distress descended upon a monkey which was blasted with air upon making errors during a difficult learning problem. We built a surrogate from whose body blasts of high pressured air blew the hair flat on the infant's head and body. We next concocted a mother who shook until the baby's teeth chattered, and finally one far removed from the concept of contact comfort, a surrogate covered like a porcupine with brass spikes. Each time, in turn, the infant indicated distress by crying and screaming, but until repulsed by the solid spikes, clung even more tightly to the mother's body. Even while clasping and rocking themselves, away from the iron maiden, the little rhesus peered out of the corners of their eyes until the spikes were retracted, then rushed rapidly to reembrace the beloved body. We had succeeded in bringing ourselves close to neuroses, but the baby did not succumb. It just returned from temporary tribulations to love mother more.

Fig. 97. Surrogate stimulation of infant love.

Prior to our planned entry into the attempted production of psychopathic behaviors in monkeys, we created an utterly unplanned succession of abnormal animals. During the first years of the Wisconsin Primate Laboratory, we worked with imported, feral animals. Pregnancies were rare and resulted only when the urge to return to the wild caused animals to escape from the cages. However, to counteract the high incidence of disease in our imported rhesus, we decided to enter the baby business and did so as soon as we moved from our small laboratory on the wrong side of the tracks into a fine, newly renovated former cheese factory. There was plenty of room for an official breeding colony of mothers and babes. The first 47 babies were carefully separated from the mothers at birth, hand-fed, and then housed in individual wire cages where they could see and hear their roommates but not intermingle or make contact with them. This system permitted us to study the infants without risking loss of limb, if not life, in prying the baby away from mother each and every day.

As month after month passed, the infants thrived but we observed that they were displaying more and more bizarre behavior. They exhibited an exorbitant amount of thumb or toe sucking, would clutch themselves and huddle, or rock back and forth in rhythmic repetition. Many would sit staring straight ahead, paying no attention to people or peers. Because of a housing shortage, some of these monkeys were placed in pairs as they reached sexual maturity, but they treated each other like brother and sister, living with propriety in perfect propinquity. As the animals matured, we paired all of them and awaited the healthy offspring we had confidently counted on. Year after year passed and we realized there was less and less likelihood of our producing a breeding colony by these methods. Where creative research had failed, we had by accident produced a plethora of neurotic monkeys. It became clear that early deprivation of social interaction was an enormously effective procedure for the development of psychopathological behavior patterns.

There are many advantages in using human subjects in psychopathological researches and many advantages in using simian or other nonhuman subjects. These subject pools were not given by God to confuse and confound psychiatric investigations. The animals and the data they provide are not contradictory, but complementary, to the human data and should be treated as such.

However, since it is, for the most part, ethically impossible to perform the manipulations necessary to delineate the factors involved in the etiology of human psychopathology, many researchers have turned with perfect propriety to the study of the development of psychopathology in nonhuman primates. The study of simian psychopathology involves the manipulation of possible etiological factors, the characterization of the psychopathology that is induced by these factors, and comparison of the behavioral changes in the monkey with behavioral aberrations in the human. Even though there are limitations to the use of nonhuman subjects in psychopathological research, the limitations are not as loathsome as indicated by Kubie (1953) who did not believe that subhuman primates were suitable subjects for psychopathological research. This is Kubie's absolute right, but we hope that he is absolutely wrong.

As pointed out by Mckinney and Bunny (1969), three primary criteria must always be met in judging whether or not a human psychiatric syndrome has been produced in subhuman animals, either tenuously or in totality. The external behavioral expression, and hopefully the internal expression also, must resemble or mirror the form of the deviant human behaviors. Second, the causative agents producing the neurotic or psychotic state in the subhuman animal must be equivalent to, or identical with, those producing the human psychiatric condition. Finally, therapeutic agents or events capable of cure or containment of the disorder in man should be similarly effective on the treated subhuman forms. Of course, if no cure has as yet been found to be effective in containment of the human disorder, and some cure effective with the subhuman forms proves to be helpful with the human forms, that might be considered to meet the criterion.*

*For further detailed discussion the reader is referred to Harlow, H. F., & Novak, M. A. Psychopathological perspectives. *Perspectives in Biology and Medicine*, **16**, 461-478. Copyright 1973 by The University of Chicago Press, from which large excerpts have been reprinted by permission.

CHAPTER 15

BROKEN BONDS

There is great community of primate psychopathology between various and divergent species. One should not expect all human psychopathology to be mirrored in monkeys, though, because these are two totally different primate forms. However, when one seeks to find analogies, one is well informed to choose infant forms before they have begun to diversify into a multitude of adult patterns and before learning, including cultural learning, has shaped them in specific but proliferated patterns.

We started to study monkey psychopathology by studying human infant psychopathology models. We started with depression and tried to produce infantile depression using as our human model the well-known syndrome of anaclitic depression described by Bowlby and Spitz. An example of one advantage of studying monkeys is that in the human being the only separation phenomenon you see normally and naturally is mother-infant separation, but this is only one of the forms of anaclitic depression. With monkeys we have been able to show that you can separate agemate friends of the same sex or of different sexes, as you choose, and still obtain an analogue of human anaclitic depression.

Anaclitic depression in human children refers primarily to the psychopathic condition resulting from the separation of mother and infant. Such separation involves the disruption of already established affectional bonds between the mother and infant, in contrast to isolation which summarily prevents any formation of meaningful love relationships. In other words, separation refers to deprivation, and isolation to privation, of maternal-infant love. The complexity of human relationships produces situations, however, in which it becomes difficult to maintain this clear-cut distinction. There are children, not separated geographically, even by diminutive distances, who may be deprived of a mother's love, since they are pragmatically separated by a more insurmountable, less identifiable behavioral distance difficult to transcend. There are mothers who develop no warm feelings for their own infants, or only for the first born. There are mothers who lose affection for their first born when he ceases to be a baby and the mother's love is transferred to the baby sibling, as long as he in turn remains a baby. There are mothers who love infants but do not love babies because of the animal-like demands they make upon the mother. There are babies, particularly autistic babies, who do not spontaneously love their mothers and mothers deprived of the behavioral feedback of baby love find it difficult or impossible to love the unloving baby in return for nothing. Mother-infant separation may be built upon biology to as great or greater an extent as upon geography, since separation is as much a matter of function as of geographical fact.

There have, over centuries of time, without doubt been unlimited, unknown or unsung cases of children separated at an early age from their mothers, but the first famous chronicles concerning actual separations were descriptions by Burlingham and Freud (1942) of war-separated children placed in nurseries in England during World War II. Adult depression has sometimes been declared to be traceable to the states of anxiety and depression rampant in children separated permanently or for any length of time from the mother. There are a number of studies which show that delinquent youth have a significantly higher incidence of childhood separations from the mother than do control groups of nondeliquents. Even more disturbing are the case histories of two chilling characters, the known assassin of Austrian Empress Elizabeth in 1898 and Oswald, the accused assassin of President Kennedy in 1963, both of whose histories are of dire and damaging deprivation. There was desertion, periodic or perpetual, on the part of both mothers and constant changes from one impersonal foundling home or school to another.

War must assume responsibility for many of the most disruptive instances of maternal-infant separation. These cases include war-deserted children, evacuated or resettled children, and children confined in concentration camps. In war nurseries, the motherless children might attach themselves adamantly to favorite feminine figures among the nurses or attendants only to have terror of separation return if one of these favorites were in turn removed by marriage or other machination. The effects of human infant deprivation of maternal love are confused and counfounded by the fact that separation is often succeeded by institutional living. The ravaging results of maternal deprivation were first brought to light and subjected to study in children centers, nurseries, orphanges, and hospitals, and the term "hospitalism" has been loosely used to cover many of the ills intrinsic to the situation. Added to the loss of love was the lack of love prevalent in institutional treatment prior to scientific studies of separation.

Beyond question, the social relationship that has received the most attention from psychologists is the mother-infant dyad. Justification for this emphasis comes from ethological, philosophical, and methodological considerations. It is generally accepted that among all mammalian species, an infant's first social responses are directed toward its mother or mother-substitute. Hence, for any given infant, the relationship with the maternal object is assumed to be first chronologically, highest in probability of occurrence, and freer from effects of experience than any possible subsequent social relationship it may enter during its lifetime. Infant monkey attachment to the monkey mother has been discussed in the section on affectional systems in this book and in human literature by various authorities interested in the attachment process, developmental data, and the process of infant participation in the world of people, starting with the first and foremost influence, the mother.

There have also been a number of investigations of occasions in which this normal pattern has been interrupted—more specifically, where the infant has been separated from its mother. In the human literature, these consist primarily of observational studies of hospitalized children or orphans. Among the most spectacular of these studies are the reports of Spitz (1946) and Bowlby (1960) of behavior syndromes exhibited by children separated from their mothers early in life. Although the children observed by Spitz were between 6 months and 1 year of age, while those observed by Bowlby were considerably older, between 2 and 5 years of age, the descriptions of the subjects' behavior during the period of separation from the mother were qualitatively similar. Both reported an initial period of active protest or agitation, characterized by marked increases in such vocalizations as crying and screaming and in activity. This behavior did not persist, but instead was followed by a stage which Spitz called "withdrawal" and Bowlby termed "despair," embodying dejection, stupor, retardation of development, retardation of reaction to stimuli, marked decrease in activity, and withdrawal from the environment (Fig. 98). It was these behaviors, along with the physiognomic expression which would in an adult be described as depression, which led Spitz to term the general syndrome "anaclitic depression." Spitz

Fig. 98. Four faces of monkey despair.

reported almost complete and immediate recovery in his children upon return
to their mothers, whereas Bowlby observed a third stage which he called
"detachment," involving hostility toward the mother upon reunion. This
inconsistency may have resulted from the difference in age of the children
observed by the two authors. Bowlby's studies set a standard for systematic,
scientific appraisal of the problem of institutionalized children and stimulated
research in the field. He made a strong case for radical reform in the care of
children raised away from their own families.

Human mother-infant separation is a subject not easily experimentally
explored. The variables involved are virtually uncontrollable. The age and
environment of the subjects as well as the length and determinants of the
separation are factors not freely decided by the investigator who has only
available cases with which to work. The physical condition precipitating the
hospitalization of the child confuses and confounds separation effects. The
attitudes of all caretakers, parents, hospital and orphanage personnel may
influence the investigations allowed the experimenter. Ethical considerations

virtually eliminate systematic manipulation and it is not surprising that scientists have turned to other species for subjects.

Bowlby concluded that the anxiety resulting from the separation of the mother and infant was produced by disruption of biologically basic mechanisms which were quite similar to the variables producing the infant monkey's affectional bonds with the monkey mother. Harlow attached more importance to clinging and Bowlby more to sucking satisfaction, but otherwise the mechanisms are comparable.

There is no doubt that clinging, which produces contact comfort, is a stronger and more persistent variable than nursing for the monkey. The human infant may first obtain his contact comfort from cuddling rather than clinging, but any such differences in the importance of components for man and monkey are probably a function of the monkey's maturational acceleration. Visual and auditory awareness and responsiveness to the mother have also been identified in studies concerning infant attachment of both species. We have found these responses both earlier and stronger in the monkey than Bowlby reports them in the human infant, again probably because of the earlier maturation of rhesus' responses.

Techniques developed in the latter half of the 20th century, however, have made possible refined and accurate measurement of the visual and auditory responses of the human infant. The neonate has been newly discovered to be a participant in many interactions long acknowledged by only the mother, whose testimony was considered to be definitely biased. Eye-to-eye contact with the mother is now known to play a part in maternal-infant communication as early as the first week and probably as early as the first day of life. Auditory reactions are more thoroughly organized than could have been ever guessed a few years ago. The very young baby reacts to the mother as an integrated whole and is emotionally upset if mother's voice suddenly comes from a different direction than where he sees her to be.

At the Wisconsin Primate Laboratory, hundreds of rhesus infants have been separated from their mothers in the first 12 hours of life. In practically all cases masculine might is required to restrain the mother during the removal of the baby. Although the emotional reactions of the mothers are brief and without residual animosity, the accomplishment does not lack hazards.

Our first, informal investigations of infant-mother separations were made with 30- to 90-day-old infant rhesus raised with their own mothers. Acute disturbance of both mother and child occurred during the separations and continued for a duration ranging from a few hours to a few days. There was a marked increase of maternal protectiveness following the return of the infant. In one instance, after a 3-day separation, the mother kept her infant literally and laterally within reach for more than a month.

After our preliminary observations, we initiated a more extensive, more

Fig. 99. Infant protest at separation from mother.

formal study. A high degree of disturbed emotional reactions characterized all mothers and infants upon separation. The infant behavior generally included high-pitched screeching, crying, and scampering, which was both disorganized and disoriented. The baby attempted to go through the plexiglass dividers and huddled against the barriers in as close proximity as possible to its mother (Fig. 99). The mother threatened and barked monkey-wise at the experimenter, but was evidently less intensely upset and for a shorter period of time than the infant. To our surprise, during the entire 3 weeks of separation, the mothers became more acceptant of the situation than did the infants. This was one of our first clues to the fact that the infants' love for the mother is of greater intensity and persistence than the love of the mother for the child, and we do not underestimate maternal love. The emotional disturbance of the babies showed a decrease as the weeks passed, but for the first half of the period was extreme (Fig. 100). Watching through the glass barrier, they kept their gaze on monther and crying continued during the same time interval.

Complex social behaviors do not well survive emotional disturbance. Play does not proceed as usual either in a strange environment or in an unhappy situation, and separation was unquestionably not a happy happenstance. In our pre-separation playpen apparatus, as in our playpen studies generally, the

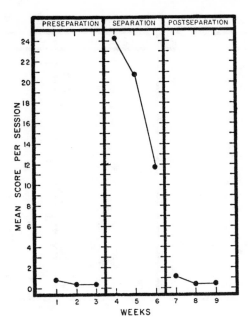

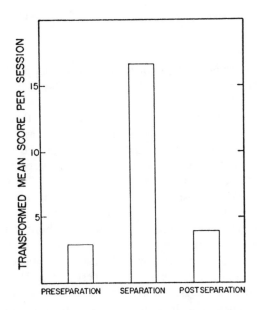

Fig. 100. Increased disturbance behavior during separation: protest call (top) and crying (bottom).

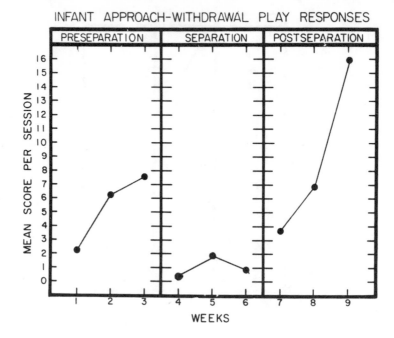

Fig. 101. Depression of play responses during separation.

infants took every opportunity prior to separation from the mother to engage in play. The drastic decline and depression of play during the separation period is shown in Figure 101.

Individual case histories of children separated from their mothers during the first years of life are replete with statements such as, "The boy did not initiate play behavior with other children," "The girl did not know how to play," or "The most disruptive effect of maternal separation was on the socialization of the child." Langmeier and Matejcek (1976) selected the most cogent conclusions from many reports on children arriving at institutions in Czechoslovakia after incarceration in the Terezin concentration camp. Psychological damage showed itself very quickly in most of the children and persisted long after general physical health was quite thoroughly recovered. Upon arrival, the young children were irrationally irritable, often aimlessly running around, screaming, tearing, and destroying. They were developmentally retarded in all areas, but especially and most severely in their social behaviors. They knew not how to eat, how to play, or how to keep themselves clear and sanitary. Other abnormal emotional responses were panic at the sight of animals and constant fear that someone would take their toys or food from them. There was striking distrust of the strange people and the strange environment surrounding them.

Since the formal separation studies (Seay, Hansen, & Harlow, 1962), extensive research on the disruption of the strong infant-maternal attachment bonds has been conducted by ethologists, psychologists, and psychiatrists. The species covered have extended over as wide a range as that of the investigators and has included avian, ungulate, canine, and human subjects. The macaque has been the most widely studied genera of monkeys. Invariably there has been evidence for the existence of the stage of protest with agitation, increased vocalization, and locomotion. In most macaques this is followed by despair with attendant inactivity, withdrawal and self-directed passive behaviors followed by the final stage of maternal reattachment.

In an interesting separation study of Rosenblum and Kaufman (1968), pigtail macaque and bonnet macaque monkey mothers were removed from their infants and from the social groups involved. Previous studies by the same investigators had shown marked differences between the natural mother-child relationships. Pigtail mothers, like the rhesus, were possessive and exclusive with their babies, whereas the permissive maternal bonnet allowed her babies to wander freely and relate in friendly fashion with other adult females who reciprocated the affection. When the mother was removed from the social scene, the pigtail infants reacted true to the expected macaque form with the first two stages of the separation syndrome, protest and despair. The bonnet babies, on the other hand, exhibited initial protest but within 2 days had found their own surrogates, sometimes just reinstating an already advanced social relationship. Unfortunately, Kaufman and Rosenblum did not conduct any subsequent major research in which the infants were separated from the mothers to establish or qualify the infants' immunity to separation despair.

Primarily interested in clarifying various separation parameters, several British investigators (Hinde, Spencer-Booth, & Bruce, 1966) subjected rhesus monkeys to, first, a 6-day maternal separation and, next, to two separate 6-day separation periods, and, finally, to a 13-day period of maternal-infant separation. The age differences of the subjects fell between 21 and 32 weeks and results showed no significant differences in separation reactions at different ages. Male monkeys were more distressed than females, and infants rejected more frequently by their mothers before separation were more severaly affected by the separation. Consistent with Wisconsin investigations, the longer the duration of the separation the more severe the distress behaviors.

In most of the studies on monkeys, separation had been limited to the age range around the middle of the first year of life until we permanently separated rhesus monkeys from their mothers at 60, 90, and 120 days (Suomi, Collins, & Harlow, 1973). They were subsequently housed either individually or in pairs. Although subjects' behaviors did not differ appreciably prior to maternal separation, this experiment revealed significant differences related to age and housing variables following separation. In general, subjects separated at 90 days of age exhibited a more severe immediate reaction to separation than other subjects, although this difference was not evident on long-term measures. Animals housed singly after separation showed far more severe reactions, both immediate and long term, than did animals housed in pairs. The 60-day-old babies were taken from their mothers prior to maturation of the fear response, which normally occurs from 70–90 days of age in the rhesus. The increase in response of the 90-day-olds upon separation agrees with previous findings that gross environmental changes produce greater adverse effects immediately following the maturation of fear.

Subjects housed without cagemates tended to exhibit higher levels of disturbed behaviors, especially higher levels of such self-directed behavior as self-clasp and lower levels of locomotor activity. The social stimulation provided either more solace and security or more stimulation to look up and locomote. Either way, roommates produced an improvement in adaptation. This housing effect was apparent to an even greater extent over a long period of time. By 6 months of age, the social scores of the separated but paired infants compared favorably with those of well-socialized, nonseparated control subjects. This result could be positively attributed to the social situation of the cagemates or negatively to the effects of being housed in individual wire cages, depending on which side of the coin is observed.

Unfortunately, there are a number of factors inherent in the infant-mother separation experiments which are less than desirable. First and foremost is the very act of physically separating a monkey infant from its mother. It is, at the very least, a most traumatic and exhausting procedure for both subjects and experimenters. The mother, holding the infant, generally requires two handlers to carefully restrain her while a third gently but firmly eases the infant from her grasp. This act, in itself, probably contributes to the subsequent agitation exhibited by both mother and infant and is not one to be repeated, unless absolutely necessary, to prevent future greater trauma in unavoidable life situations.

Suomi (Suomi, Harlow, & Domek, 1970) later achieved an alternative approach to the production of induced depression via social separation of infant monkeys. He separated agemates or peers rather than mother and child and measured the effects of multiple repetitive separations and reunions rather than a single separation and reunion.

The model of mother-infant separation undoubtedly was influenced by Freudian theory which hypothesized that mother love was both the primary and the all-pervasive affectional system in human beings. We questioned this position, without denying priority or potency of the maternal role. We suggested, in rather forceful terms, that the agemate or peer affectional system might be of even greater importance to the development of subsequent social and sexual behaviors. Even if maternal love has chronological priority, agemate love is temporally more tied to the subsequently learned integration of adolescent and adult behaviors.

Limited investigations involving multiple separations of monkey mother and child can be found, but they do not have the scope of the agemate repetitive separation studies. The primary contribution of these studies was in demonstrating that multiple infant-mother separations were technically possible.

The Suomi study utilized brief, multiple separations to determine the effects of loss of agemate love, rather than maternal love. At 90 days of age, a group of four monkeys, raised together without mothers, were subjected to a series of 12 separations, each of 4-day duration. After a 6-week period of group housing, they experienced a second sequence of eight 4-day separations.

The infant monkeys exhibited a typical period of protest during each of these agemate separations. Intense vocalizations and attempts to reach the separated partners during separation did not extinguish within a 6-month period. Separation produced prolonged pain and persistent protest. Protest in each of the 20 separations was followed by a period of despair, characterized by decreases in locomotion and increases in self-clasping and huddling.

The most surprising finding of all was that the infantile behavior characteristic of any and all 90-day-old monkeys persisted throughout the half-year study. At the beginning of the study the subjects were at an age where basic infantile partner clinging constituted the primary social behavior. This activity rapidly disappears in normal monkeys, but for the entire study the infantile behaviors, particularly clinging, persisted and the social behaviors of exploration and social play never did burgeon into full flower. At 9 months of age, the multiply-separated monkeys still clung at 90-day levels (Fig. 102). They failed to display any indication of development of complex social behavior, particularly play, that normally dominates the social life of macaque monkeys from 4 to 8 months of age (Fig. 103). If one gives full weight to the persistence of infantile primate behaviors at 9 months of age and the picture of the paucity of social exploration and play at the same time, one can only assume that the multiple separations maintained the complete infantile stage of the 3-month-old macaque mind and motives. Here infantilism was not produced by regression but was continued by maturational arrest. It is apparent that the technique of multiple peer separation was at least as effective in producing anaclitic depressive behavior as the more traditional maternal separation procedure.

The persistence and strength of the depressive behaviors of the separated together-together raised monkeys contribute convincing evidence of the power

Fig. 102. Infantile together-together clinging.

of peer attraction. This affection, if not stronger at this time of life, is at least as binding as the infant-maternal tie. As time passes, the influence of the maternal bond decreases and rhesus infants, just like their human counterparts, choose agemate association in preference to maternal shelter.

Peer separation has not been as readily studied in human life as has infant-maternal separation. The closest comparison which can be made at present is in the effect of loss of siblings, and even here any real experimental evidence is lacking. Langmeier and Matejcek (1976) report a long-term separation of a 5-year-old from his younger brother. This case was complicated by the death of the mother and desertion of the father. After

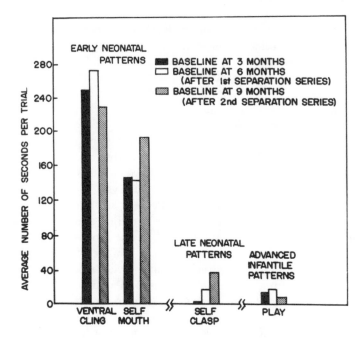

Fig. 103. Persistence of infantile self-directed behavior and paucity of play development in peer separations.

two years with the grandmother, both boys were sent to different children's homes, and the older brother returned for a visit to discover that his younger brother had been adopted and moved many miles away. Runaway behavior ensued for a number of years and even after the older brother had eventually become outwardly reconciled to the children's home he talked about his brother whom he loved more than anyone else.

Other human children who may be considered products or, perhaps more pertinently, casualties of separation situations, are children who react to the acquisition of a new baby in the family and suffer all the symptoms of desertion and loneliness. These cases can confound the best of clinical authorities. In one family the mother gave very intelligent and equally loving attention to the second child who was just a year and a half when her new brother arrived. The child liked this increased attention so much that she played her part to the hilt, demanding attention and crowing for more care. Suddenly and almost belatedly the mother realized what the combination of both baby and brat was doing to the oldest child (4 years of age) and quickly began regrouping her forces.

There have been a number of experimental attempts at Wisconsin to produce depressive behavior in older monkeys, primarily in adolescent monkeys. The more successful of these studies have been conducted with subjects reared exclusively in nuclear families. The nuclear family situation, designed by M. K. Harlow, consisted of units containing four mother-father pairs and their offspring. The infants had free access to all adults and peers.

Ten 5-year-old offspring from the nuclear family were removed from their families. Four of these, all from one family unit, two males and two females, were housed together. Four subjects, two each from two other units, were also housed together, and the remaining two subjects were individually housed. This arrangement was maintained for 60 days.

Five-year-old monkeys approach full-fledged adulthood. Therefore, no matter what the results, we would not expect to find anaclitic, or childhood, depression. What we did find was that all of the group-housed adolescents suffered only a mild reaction to the separation from their families. The groups with offspring from two different family units, in addition, showed fairly high levels of aggression from being suddenly placed in an intimate living situation with less familiar agemates and without their own families. The individually housed rhesus, however, went through the equivalent of the first two stages of anaclitic depression, which, for lack of a better name, will be given the human adult equivalent, the nondescript name of plain depression. Initial extensive agitation upon separation was followed by deep depression.

We might be tempted to judge that separation of these adolescents produced depression because the environment of the nuclear family unit produces more meaningful relationships among offspring of this age group than have the environments of other separated adolescents. However, although these studies show that syndromes analagous to human depression may be produced in monkeys of different age groups, we have also been cautioned, from past hisgory, not to judge from results of single studies. By no means do the results of these studies show that every social separation will yield depressive reactions. Suomi produced depression by separating younger, intimately associated peers from each other, but in a later study we found no such results from a separation of older, less well-attached peers.

What we have at present is a preliminary appraisal of some of the variables involved in particular separation situations. First, as was earlier mentioned, bonnet babies have less binding attachment bonds with their mothers than do macaques. They are very unlikely to exhibit profound depression upon separation, as do both the pigtail and macaque who share strong attachments and close relationships with their respective mothers.

Second, subjects who are placed, upon separation, in new environments which offer some components or satisfactory substitutes for the constitutents of the original environment are unlikely to display deep depression. Nuclear family reared semi-adults who remain with part of their family or with friends show little effect of the separation. Separated from both family and friends, however, and not allowed any species-specific interaction, nuclear family semi-adults respond with deep depression. Being a representative of a nuclear family, by itself, does not produce immunity to depressive reactions. Also, the presence of peer partners upon separation allows social development to proceed upon schedule. Individual housing upon separation not only affects subsequent social development but produces further deleterious depressive effects.

Fortunately, the reaction of both children and adults to the majority of separations is mild and transitory. We believe that the proper assessment of the variables examined through these studies may appreciably facilitate accurate prediction of the occurrence and severity of monkey response to separation and facilitate, as well, understanding of separation in human beings when depressions do occur, as we know they do.*

---

*For further experimental data and details the reader is referred to Seay, B., Hansen, E., and Harlow, H. F. (1962). Mother-Infant Separation in Monkeys. *Journal of Child Psychology and Psychiatry*, **3**, 123–132.

CHAPTER 16

THE HELL OF LONELINESS

Love is by most definitions a social phenomenon. When you prevent any social situations from ever occurring, you obviously prevent any vestige of love from ever developing. Technically this phenomenon is called social privation, which includes human social isolation of all degrees.

Human social isolation is a problem of vast significance and of varied incidence. Social isolation may be of a partial nature, such as might occur during periods of brief breakdowns of the family structure due to illness. Members of the family may each experience partial isolation effects in entirely different manners. The mother may be hospitalized and as a result the child or children dispersed to the homes of friends or relatives, leaving the father a bachelor's existence, to burn his toast all by himself. To each his own, but in relatively strange situations, strange in different degrees.

Or death may cause a more permanently disastrous disruption, sending the offspring to an orphanage or foster home. The problem of such partial isolation becomes entwined with the additional sadness and terror of separation and loss of loved ones.

The extremes of human social isolation might be, at one extreme, a child's first day at school and, at the other, the solitary confinement of a criminal offender. The strangeness of a child's first day at nursery school, kindergarten, or first grade, after mother leaves, will usually dissipate after the first few days among socially raised tots. The total social isolation of solitary incarceration is considered so drastic that Americans pride themselves on reserving it for the most pernicious prisoners.

Leaving familiar friends and family, going away to boarding school or college, or being transferred in business to a new city may be fearsome and formidable forms of partial isolation.

One of the authors, alone, travelled two thousand miles west to college. The first words she remembers hearing and the first words she remembers saying came about two weeks after arrival. "I'm from Minneapolis," said one of her dormitory mates. "You're from *Minneapolis!*" cried your author. "I used to live there." I had moved away from Minneapolis years before, but, at that moment the familiar place meant home.

Your other author, fresh out of graduate school, drove two thousand miles east to assume—assume, if not the correct, is the intended word—his first teaching position. He was to stay at the University Faculty Club and proceeded to drive directly there. He went to the desk and said, "I believe you have a room for Dr. Harlow?" "Yes," came the reply. "Do you know when he is arriving?"

Still shaking, the very new Dr. Harlow went up the hill, you might say over the hill, to his office and sat down at his desk. A few minutes later, a breathless young man about Dr. Harlow's age rushed in and said, "Do you know where I can find Dr. Harlow?"

Partial social isolation refers to isolation from the members of one's own species, be it monkey or human, as well as lack of interaction with others, but it does not preclude seeing or even hearing these conspecific individuals.

Total social isolation, as its name implies, means complete prohibition of both visual and auditory exposure to others, as well as the total absence of interaction with other individuals of one's own species. Since Homo sapiens is the only living species of the genus man, total social isolation infers isolation from the whole wide world of man.

If the data were true, extreme total human social isolation might be illustrated by the children of India and the children of Africa who were raised by wolves or other beasts in the wild. Unfortunately, the stories of most of the wolf children proved to be prefabricated, and the age of isolation unknown. We also have no confirmation, one way or another, on the parentage of Romulus and Remus. The story most beautifully and convincingly documented, of the wolf children of Midnapore, India (Singh and Zingg, 1966), although missing the times of parental separation or desertion, is primarily the story of separation from the feral family rather than from human, or inhuman, relatives.

Much more common cases of total social isolation, however, are but too true. These are the cases of political, wartime, or criminally dangerous prisoners who are placed in solitary confinement in present day prisons all over the world and not in dungeons of old like Kasper Hauser. Kasper was a Bavarian prince who was confined in a single-room dungeon all by himself as a very young child and denied all human contact until shortly before maturity. This is a tale which is loosely related to fact, but many of the actual facts are missing. Some years later, Kasper was also still missing.

Not only is it difficult, but it is also undesirable to study scientifically the impacts of culturally produced total social isolation at the human level. The variables are multitudinous and recalcitrant to experimental manipulation and control. The subjects are even more recalcitrant as well as reluctant to be experimentally manipulated. Many are quite uncontrolled, especially after any type of social isolation.

Fig. 104. Partial isolation cage.

If you have ever witnessed any of the tragic traumas which social isolation may produce, you must realize the vast importance of the problem of human isolation. Total social isolation is not confined to criminals, unfortunately, on the human level. Abusive parents or unscrupulous baby tenders are known to have locked little children in dark closets, alone, or tied them to attic bedsteads far from any madding crowd or other welcome noise.

We first encountered the destructiveness of partial social isolation when we raised our rhesus monkeys in individual wire cages (see Fig. 104) in order to curtail the disastrous, contagious diseases brought with the monkeys imported from India. At that time, the importance of peer love and agemate play was nowhere near fully realized. Indeed, many facets of the value of agemate affection and peer play became realized through the inadvertent incidence of raising the monkeys in partial social isolation. Knowledge of the extent of maternal contributions to infant weal and welfare was also unexpectedly increased through the same experience.

Cross and Harlow (1965) reported the syndrome of compulsive behaviors which become ever more severe as partial isolation is prolonged. These maladaptive behaviors include nonnutritional sucking which serves no ordinary purpose, stereotyped circular pacing, fixed and frozen bizarre bodily postures and positions of hand and arm, as well as withdrawal from the environment to the point of complete detachment.

While both ethical and practical considerations complicate the controlled scientific investigation of total social isolation with human subjects, we have been able to systematically study total social isolation with the closely related, nonhuman rhesus monkey primates.

The overall effect of social isolation on the primate is a function of both the length of the isolation period and the stage of development at which isolation commences. For instance, monkeys totally socially isolated for 12 months from birth on show a greater frequency of withdrawal into self-directed, bizarre behaviors than do monkeys totally isolated for the first 6 months of life (see Fig. 105). Animals totally isolated after some social interaction with mothers or peers do not reach the depth of behavioral deterioration which envelops animals isolated from birth for the same time periods. If a baby monkey has had the experience of mother love first, there is a residual reaction. Even better, if the baby has been allowed to partake a little longer of monkey society and share play with peers, the effect of the incarceration will again diminish.

We conducted a series of experiments in which we housed monkeys from a few hours after birth until 3, 6, or 12 months of age in a stainless steel chamber. During the period of time in the isolation cage, the monkey had no contact with any living animal, including the experimenter. We purposely made no pretense of creating any sensory deprivation of sound or light. We

Fig. 105. Twelve-month total socially isolated rhesus monkey
showing extreme withdrawal and bizarre posture.

also did not eliminate opportunities for simple exploration and motor activity.
We wished the lack to be limited to the social being.

Upon exposure, after 3 months of isolation, to any part of the great wide
world outside their cage, the monkey infants go into a state indicating panic
or shock (see Fig. 106). We have found throughout the years that monkey
infants at this particular age are extremely vulnerable to formidable fears.
Since fear of the new and the strange is an innate fear, it is no wonder that
the isolates react so drastically. Every single object, whether another monkey
or a play toy, beyond the isolation cage is completely new and strange.

Fig. 106. Three-month old rhesus just removed from the isolation cage.

Isolation effects were measured by comparing the social behavior of pairs of isolated monkeys after the chambering experience with that of pairs of equal-aged monkeys not confined in the isolation chamber. Each experimental pair was tested in the playroom with its control pair (see Fig. 107).

The debilitating effects of 3 months of total social isolation are dramatic, but seem to be completely reversible. If there is any long-term social or intellectual damage, it completely eluded our attempts to find it. Given the opportunity soon after release to associate with controls of the same age, these short-term isolates start slowly during the first week and then adapt and show the normal sequences of social behaviors. In human terms, they are the children salvaged from the orphange or inadequate home within the first year of life.

The unfortunately umambiguous findings of numerous researches on rhesus monkeys totally isolated for at least the first 6 months of life have told a different story (see Fig. 108). Total social isolation for these first 6 months enormously damages or even destroys the normal, subsequent social and

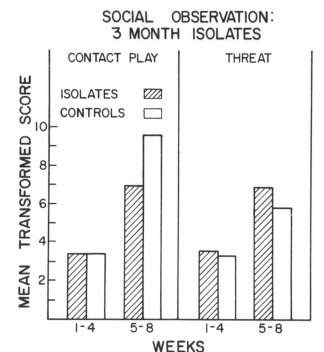

Fig. 107. Deficit in play development of 3-month totally socially isolated infants.

sexual capabilities. The infant monkeys were inevitably inept and incompetent in their interactions with socially normal agemates. Normally, by 6 months mother and peer raised rhesus monkeys have developed complex social patterns including vigorous play and socially protective aggression. Any stranger introduced into their group will be attacked and will be allowed to join the group play only if he fights back. If he is not reciprocally aggressive, the strange monkey will remain the victim of the aggression of the original members of the group.

The behavior of the isolates upon release from isolation is so different from that of the average rhesus of 6 months of age, it is not surprising at all that it should be taken for a stranger even without invasion of an established group. Nor is it surprising that recovery should be thus impeded. Upon emergence from 6 months of total social isolation, the behavioral repertoire of the young monkeys is dominated by such infantile self-directed activities as self-clasping, self-mouthing, huddling and rocking repetitively back and forth (Fig. 109). Such behaviors make social behaviors not only unlikely, but completely

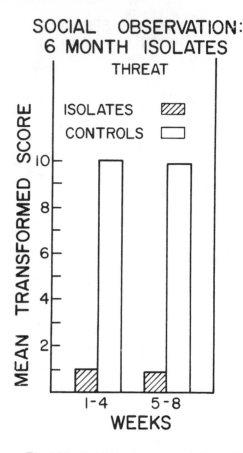

Fig. 108. Total inadequacy of 6-month
isolate in socially protective threat responses.

incompatible. The 6-month isolates failed to initiate or reciprocate the exploration, play, and grooming enjoyed by their normal, socially raised peers (Fig. 110).

The difference between the 3- and 6-month isolates showed markedly in the most significant symbol of social development, play. Contact or rough-and-tumble play showed practically no difference between the 3-month isolate and the normal control animals, either immediately after isolation or in the subsequent developmental trend. On the other hand, for the 6-month isolates there was essentially no play for 8 weeks of post-isolation periods in the playroom (Fig. 111). Interactive play between the isolates developed somewhat by the end of 6 whole post-isolation months, but even after 8

ISOLATION PERIOD

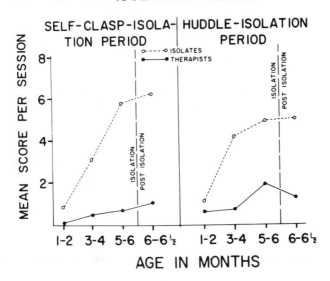

Fig. 109. Infantile, self-directed maladaptive behaviors of 6-month isolate.

months social interaction between the controls and the isolates was still nonexistent, except for intermittent bursts of aggression by the controls toward the isolate animals.

The effects of 6 months of social isolation from birth on adversely affected the monkeys for the rest of their lives. Three years after the 8 months reported in the playroom tests with the normal peers, the same isolates were tested again and their repertoire of behavior had deteriorated even further, without a trace of improvement. Their sexual behavior lacked all but a semblance of sex, and that with possibly good but obviously misdirected intentions. The only evidence of maturing was in an increase in reactions of fear and aggression, performed just as inappropriately as their sexual sorties. They aggressed against infants as no normal monkey would deem decent, and several isolates actually attempted single, suicidal aggressions against large adult males.

Surprisingly, the total social isolate rhesus still performs the necessary routine behaviors to sustain life and health, if not the pursuit of happiness. It eats and drinks sufficient amounts to maintain weight without interrupting the pious posture of deep depression. No monkey has died during isolation. A very occasional monkey, when removed from total isolation at 3 months of

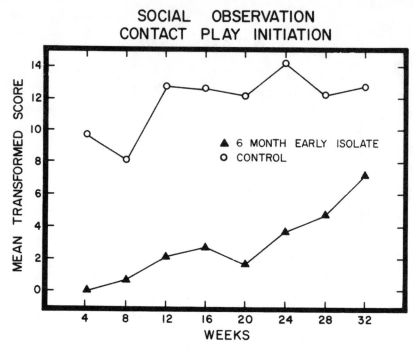

Fig. 110. Play initiation failure in 6-month isolate.

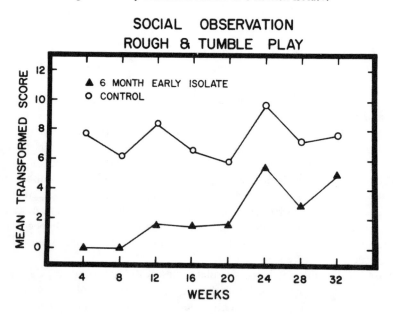

Fig. 111. Persistence of isolation effects preventing normal play development.

SOCIAL   OBSERVATION
ACTIVITY   PLAY

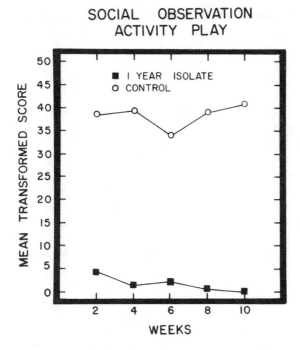

Fig. 112. Almost total lack of play activity after
12 months of isolation.

age, has shown symptoms other than depression. One died of self-induced
anorexia, and another would probably have died except we were, by then,
forewarned and we saved him by the technique of forced feeding.

With 6 months of social isolation producing such a plethora of pitiful,
maladaptive behaviors, we had originally assumed that 12 months of isolation
could not be more catastrophic. This assumption was optimistic. Twelve
months of social isolation almost obliterated the primates as social animals, as
is shown in the comparison of play of the isolates with the control animals
(Fig. 112). If practically no play 2 weeks after removal from isolation could
decrease, and it did, 10 weeks past solitary confinement the isolates had no
play at all. The controls continued regularly to show high play levels. The
experiment had to be terminated after 10 weeks of observation. The normal
animals became so agonizingly aggressive toward the isolates that we were
afraid they would terminate the experiment had we not.

Total social isolation during the second half year of life, after social raising
for the first 6 months, indicates that a little learning, like a little love, is far
better than none, and especially if that little learning is social in nature and 6

months long. Within a relatively short time, the 6-month isolates, termed the late isolates, adjusted adequately toward their equal age playmates. In other words, if you must deprive a child of social contacts, wait until you have first given the infant an auspicious launching into the wonderful world of mothers, men, and merry playmates. If you reverse the procedure, the loneliness of fright may freeze and force the fugitive back into the lonely lair. The findings of the various isolation studies with the monkeys indicate that severe and steady early isolation produces the primary social response of fear. Placed in a feral, free-living environment these animals would be driven away or eliminated prior to any period of grace to allow them to adapt to the group.

At this point, it is pertinent to remember that the early socialization of the macaque is greatly stimulated by the mother, who provides the infant with the stable source of security and trust against the strangeness of the world away from mother. This type of social security develops during the first 6 to 8 months of life in the adequately mothered human child and during the first 2 to 4 months in the monkey development. This basic sense of security encourages the infant to explore all aspects of its physical surroundings and to begin to interact with agemates, if they are present.

The 6-month isolate, raised without benefit of maternal security, is deficient in comparison to the socially raised control animals on inanimate object exploration and manipulation even as long as 20 weeks after being placed in a social playroom with the social monkeys.

We have purposely left to last the detailed discussion of partial isolation experiments conducted at the Wisconsin laboratory, since the picture has some features quite unique in comparison with any and all of the other isolation studies. By far the most common cases of partial isolation are the literally thousands of rhesus monkeys in medical and social science primate laboratories widely separated in various parts of the world, if these monkeys have been raised in a modified form of partial isolation, i.e., housed in individual bare wire cages where they could see and hear other monkeys, but not be with them or interact with them. These monkeys develop few age or sex appropriate social responses and many inappropriate, abnormal disturbance behaviors during the first year of life.

Experimental studies at Wisconsin show that monkeys partially isolated in individual wire cages, for either 1 or 3 years of life, in comparison with socially raised monkeys, exhibit many more self-directed activities, such as mouthing and clutching their own bodies, and lower levels of oral manipulation of objects and use of social threat, a behavior not as threatening as it sounds. Older monkeys, partially isolated for 1 to 2 years, had a marked decrease in the self-directed abnormal behaviors, with the exception of one of the most abusive, self-aggression, which increased and revealed the persistent effect of the asocial rearing condition.

Fig. 113. A long-term partial isolate with the vacant, fixated stare of the catatonic.

In a study of older rhesus monkeys, 10 to 14 years of age, Suomi (Suomi, Harlow, & Kimball, 1971) used as controls similar-age monkeys with an early feral life up to 2 years of age, prior to the subsequent partial isolation in the bare wire laboratory cages. The rhesus who had never known life other than that of cage confinement showed lower activity levels overall than the younger researched partial isolates. Rather than actively displaying bizarre behaviors, they primarily just passively sat and stared. Even so, they could be differentiated from the feral-born control monkeys in both lowered activity and in their exhibition of disturbance behavior patterns.

Although monkey or human children raised in partial isolation do not show normal behavioral development as we like to think of it, there are certain changes within the abnormal responses which reflect the changing physical age during each subsequent year of partial isolation. During the monkey's first year, the infantile self-directed behaviors most prominent are self-mouthing, self-clasping, rocking back and forth, and huddling. The rhesus infant follows these in the second year with lackadaisical passivity and stereotypic motions, reminiscent of the methodical, repetitive pacing of a lion caged long alone in a zoo. Since anger and aggression are late maturing behavior patterns, self-biting is one of the last activities to appear, after the second year in the males and even later in female monkeys.

We have studied many animals that were raised in partial social isolation, during which the monkeys may see and hear other animals but can not make physical contact. Some, but by no means all, of these monkeys have developed symptoms simulating catatonic schizophrenia in human beings. As they mature, they may sit almost fixedly at the front of their chamber staring vacantly into space, paying no attention to either people or other monkeys. They frequently developed bizarre stereotyped postures and twice we have seen a chilling catatonic characteristic (see Fig. 113). While the monkey sits immobile, an arm may raise; and when he sees it, he either ignores it or freezes in fright. Needless to say, this is a disarming phenomenon. In these animals, as the arm raised, the elbow, arm, and fingers flex and, if this is not a catatonic gesture, America's greatest schizophrenic authority is totally wrong. I was reporting on depression at a meeting of the National Academy of Sciences and by accident brought the wrong slides. These were supposedly slides of depressed monkeys. Knowing the automatic absorption of my colleagues, I was neither unduly surprised nor disturbed, but Dr. Seymour Kety was in the audience. An hour later, we were attending a small group meeting and he asked me how I could possibly describe these animals as depressed when they were obviously schizoid. I explained to Dr. Kety openly, at a rash moment, that I had brought the wrong slides, but I believed I could bluff without doing psychological harm. I had not figured on Dr. Kety!

It is an open question as to whether or not all forms of human psychopathology can be reproduced in monkeys. Certainly not in name, but there is enough community of behavior to justify the monkeys' phyletic position close to man.

Many related problems remain to be done. If we can replicate four human conditions, more can be achieved. We have no definitive information concerning the intensity and permanence of the psychopathological states achieved. Multiple therapeutic agents, both behavioral and biochemical, remain to be explored. Having simulated the devil, and exorcised some, perhaps we can next search for the deep blue sea.*

---

*For further experimental data and details the reader is referred to Harlow, H. F., Dodsworth, R. O., & Harlow, M. K. Total Social Isolation in Monkeys. *Proceedings of the National Academy of Sciences*, 1965, **54**, 90-97.

CHAPTER 17

LOVE RESTORED

In Wisconsin primate research, one experimental series has logically led to another, or to several new studies suggested by the latest findings. Social isolation created not only depressed and desperately disturbed rhesus infants, but also disturbed experimenters with only one dictated direction in which to go. We could not leave the babies in eternal damnation and depression without every effort toward their rehabilitation.

Several unsuccessful attempts were made, such as to reverse results of isolation by painful reconditioning with aversive stimuli, or to prevent postulated emergence trauma by anticipatory adaptation during isolation. Three months of total isolation from mother and agemates produced social deficits which proved to be reversible, but isolation for the first 6 months of life and next, 12 months of isolation consistently resulted in profound and progressive play and sex disruption (Fig. 114). As infants and adolescents, play was neither initiated nor reciprocated and social grooming languished. As adults, these monkeys still exhibited abnormal sexual (see Fig. 115), aggressive, and maternal behaviors reflected even in their infants.

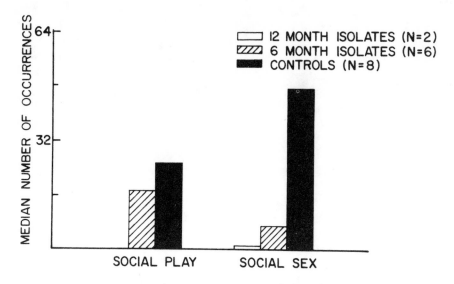

Fig. 114. Progressive deterioration of social behaviors with length of isolation.

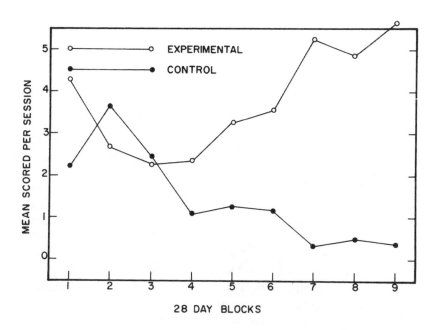

Fig. 115. Deficient maternal behavior of former isolates produces increase in self-directed abnormal behavior of their infants.

Our research on social play and peer love had come close to convincing us that peers were peerless in their powers, and the technique of rehabilitation next chosen was influenced by this attitude. We decided to socially integrate animals that had been isolated for 6 months with equal-age playmates who had been raised with normal social interaction.

Rhesus monkeys raised in total social isolation were placed in a playroom with other monkeys the same age who had been raised in respectable fashion with real mothers and peer playmates. The hoped-for transformation did not transpire, and the results were disastrous. The isolates froze with fear and were torn with terror. The isolates had no concept of play and engaged in absolutely no social interaction with their peers and almost none with each other. The isolates proved inferior on practically every positive social behavior. They showed less curiosity in exploration of objects, did not initiate approaches or contacts with the other monkeys, nor did they play, alone or with others. Disaster periodically threatened. The pitiful, pathological isolates engaged primarily in self-directed behaviors, with a show of self-clasping,

Fig. 116. Normal rhesus aggresses same-age fear-frozen isolate.

self-chewing, and huddling anti-socially in a corner (Rowland, 1964). Their behavior was so bizarre that the normal young monkeys aggressed them as if they were dangerous strangers (see Fig. 116). The primary gain from the study was the knowledge of what must *not* be done to effect any rehabilitation of isolate animals.

Toward the end of the 8 months post-isolation testing, the isolates showed some increase in play behavior, locomotion, and exploration within their own isolate group, but not toward the normal agemates who had been guilty of unprovoked aggression toward them earlier in the test period. Also, within, and only within, their own isolate group, the social threat response made its delayed and limited appearance. This response is used socially to communicate certain social roles such as dominance and submission.

Throughout the 8-month testing period, the isolates were clearly inferior to the normal controls on virtually every behavioral measure. However, their semblance of recovery and limited improvement gave the first evidence that social recovery might be a possibility, after 6 months of total, social isolation.

Fig. 117. Motherless mother ignores beseeching baby.

And little play is far better than none. Despite my gloomy predictions, a series of studies were completed and a number of others are in progress on the rehabilitation of the partially and totally isolated animals.

Social recovery of a different form was exhibited by another group of monkeys subjected to early social isolation. These were females who had reached sexual maturity but were apathetic or adamantly uncooperative when confronted with breeding-stock males that were sexually eager and adroit. By methods dark, diverse, and devious, we impregnated several of these reluctant females over a period of years. We have called them "motherless mothers," since they never experienced mother love, nor any other kind of monkey affection. Most of the motherless mothers either completely ignored or abused their initial offspring (Fig. 117). However, with unbelievable fortitude and in the face of impossible odds, the babies struggled for maternal contact day after day, week after week, month after month. The infants would cling to the mothers' backs, continually attempting to achieve ventral and breaat contact despite efforts by the mothers to displace them (Fig. 118). To our surprise, maternal brutality and indifference gradually decreased. From the fourth month onward, the persistent babies that were finally able to attain intimate physical contact with their mothers were actually punished less and permitted nipple contact more than offspring of normal mothers.

Many of the motherless mothers have had second and even third babies. Those whose maternal feelings were eventually released by the persistent and determined body contact and nursing efforts of their first infants proved to be adequate or good mothers to their subsequent babies. Most of the motherless mothers that had abused or ignored their first infants throughout a predetermined 6-month postpartum period continued to be inadequate, brutal, or lethal mothers to subsequent progeny.

It seems that in the case of the adequate mothers, the babies had inadvertently served as psychotherapists to their indifferent mothers and that these mothers spontaneously transferred maternal feelings induced by their first babies to their feelings for subsequent babies. The data on the rehabilitated mothers suggested that infants possess some specific abilities as behavioral therapists for abnormal adult females. However, rehabilitation was limited to maternal behavior.

The rehabilitative potential of social contact or contact comfort was further illustrated in a third study. Four 6-month isolate monkeys were individually housed for 2 weeks after removal from the isolation chambers. Heated cloth surrogate mothers were then introduced into their home cages. Within a few days, the isolates began to contact the surrogates with increasing frequency and duration. Correspondingly, the incidence of disturbance behavior exhibited by the isolates showed a significant decrease from presurrogate levels.

Fig. 118. Infant tries to get to the front the hard way.

After 2 weeks of exposure to the heated surrogate, the isolates were housed in pairs. To our delight, they almost immediately exhibited some social play, a suspicion of infantile sex, and a little locomotion, but our hopes were not long lasted. The social behavior was clumsy, and the isolates continued to display disturbance activity. No marked improvement materialized in 6 months.

We have long emphasized that the mother figure is a basic socializing agent, since it gives to its own infant social contact acceptability, essential to any subsequent social interaction—and basic security and trust. In this case, surrogate mothers appeared to provide a certain degree of contact acceptability and security and trust sufficient for the isolates to suppress existing self-directed disturbance activity and to initiate crude social interactions with other isolates. However, since the inanimate surrogates could provide no further social stimulation, recovery failed to progress beyond the above stage.

In all of these instances of partial recovery, the isolate animals actually gave more than they received, whether it was to each other or to the motherless mothers. The normal agemates treated them like invading strangers to be attacked, an approach not likely to encourage reciprocal love and affection. The surrogate mothers were cozy and cuddly, but again, these attributes do not provide social interchange and with the motherless mothers the contribution was one way only, a social benefit to the mother and not the child. Any thorough social rehabilitation had to be, in reality, more of an indoctrination than a rehabilitation because the isolates had never been normal social animals.

So far, all of these researches had used social agents to attempt rehabilitation. There were additional experiments using nonsocial techniques. Pratt (1969) tried exposure to slides as substitutes for the real social experience during isolation. Others introduced the principle of gradual introduction to the post-isolation testing situation (Mitchell & Clark, 1968) and aversive conditioning (Sackett, 1968). The social approach continued to suggest the strategy most likely to succeed.

The inspiration for a real social debut was found in normal, socially raised rhesus infants younger by half than the 6-month isolates. These therapists had been raised with surrogate mothers and had been, after the first month, permitted to interact for 2 hours a day in groups of two or four, in either a quad cage or in the standard playroom. The 3-month-old junior therapists, or psychiatrists, were young enough and small enough to pose little threat to bewildered isolates. We deliberately chose therapists younger than the abnormal isolates in the hope that this age relationship would ameliorate fear and flight. They were at an exploratory age when new objects, both inanimate or animate, were gingerly approached with gentle gestures.

When the therapists averaged 3 months of age and the isolates 6 months, all eight infants were placed in two quad cages, two therapists and two isolates to each cage. The isolates were very gradually introduced to the presence of each other and of the therapists. For 2 weeks the isolates could, for the first time, see and hear other animals, although still separated by mesh barriers.

Therapy, as well, followed a procedure of gradual change, gradual introduction of the isolates to the new, normal animals and to new environments. When the mesh barriers were first raised in each quad cage, one isolate and one therapist were physically free to seek each other. At first,

Fig. 119. Gentle therapist offers contact comfort.

however, the isolates each retrated to a corner where they huddled and hid, but the less timid therapists followed and carefully clung to them (see Fig. 119). At first, this social session lasted only for 2 hours, 3 days a week; then, in addition, on the other 2 days the animals were allowed, as in the ark, two by two, two therapists and two isolates together in groups of four in the regular playroom. Within a week in the home cage and within a month in the stranger environment, the playroom, the ice was broken and the isolates were clinging back. Reciprocity was revealed for the first time.

The age of 3 months had another additional advantage, since that is the age at which play begins to have an ever-increasing role and significance to developing monkey infants. We believed that if the therapist could, step by step, overcome social fears by posing no threat and subsequently gently lead the isolates into play, play which would prove pleasant and without terror, that two of the prime requisites for rehabilitation would have been achieved. The isolates, little by little, discovered the fun of frolic along with their teachers who were learning as well. The therapists spontaneously showed the very behaviors which would beguile other beginners and play began to bourgeon (see Fig. 120).

Creation of normal social interactions in play is the best criterion of rehabilitation in an immature animal. The psychotherapeutic principles of avoidance of social fears through a fear-free, gentle, and ingratiating therapist plus the inculcation of gradual step-wise social interaction will doubtless eventually be discovered by the psychiatrists. Indeed, these well-established principles of animal training have already been described by the Tinbergens.

The thoughtful psychiatrist would learn far more about the troubles and tribulations of his patients by tracing the development of play behaviors and the social functions which they subsequently serve. Play should not be considered on the premise of a disguise for individual conflicts. Play is rather useful as a diagnostic tool, to discover whether or not the patient did play, at what, with whom, how soon, and the effects upon subsequent complex social behavior or play in the guise of a background for therapy under certain conditions.

Fig. 120. Therapeutic play behaviors.

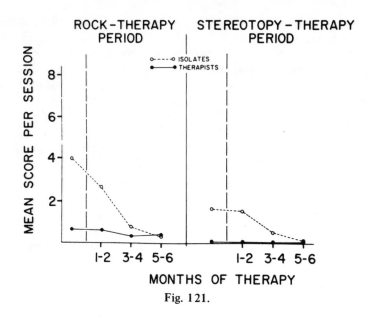

QUAD CAGE BEHAVIOR THERAPY

ROCK–THERAPY PERIOD   STEREOTOPY–THERAPY PERIOD

MEAN SCORE PER SESSION

o----o ISOLATES
•——• THERAPISTS

MONTHS OF THERAPY

Fig. 121.

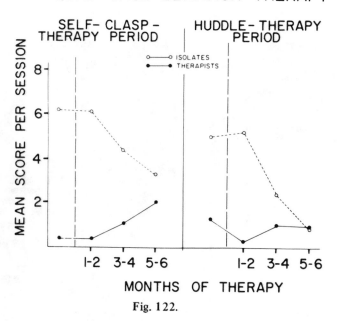

QUAD CAGE BEHAVIOR THERAPY

SELF– CLASP – THERAPY PERIOD   HUDDLE– THERAPY PERIOD

MEAN SCORE PER SESSION

o——o ISOLATES
•——• THERAPISTS

MONTHS OF THERAPY

Fig. 122.

Figs. 121 and 122. Gradual disappearance of self-directed infantile behaviors during the 6 months of therapy.

266

In the experiment with the junior therapists, play and the number of players were gradually and progressively increased. As the therapists developed more and more sophisticated play patterns, they persuaded their new playmates to perform with them and the maladaptive behaviors were discarded, to disintegrate with disuse (see Figs. 121 & 122). By one year of age, the therapists and the isolates were indistinguishable, one from the other, even to the most erudite experimenters (see Fig. 123).

Two of the most valid measures of social recovery are those of social contact and play. In the playroom, although it was large and strange, there was complete development of social contact and social play behavior of the isolate to almost the same level as that of the therapist. This is all the more remarkable in that these behavior levels are not indicative of the recovery of previously existing behavior patterns of these isolated monkeys. They were raised to these levels from the 0 level existing at the age of 6 months.

Due to the unpredictability of monkey mothers and babes, our laboratory nursery did not produce sex-balanced groups of therapists and isolate monkeys

Fig. 123. Therapist and isolate at play.

to grow to the respective ages of 6 and 3 months at exactly the right moment. We achieved four 6-month-old isolates who were all male and four 3-month-old-therapists who were all female. That the successful therapy did not depend upon this sex relationship was proven by later sex-balanced groups; but, of course, the power of a gentle, beguiling, smaller female to completely change the crude characteristics of a great giant of a domineering male has long been known.

This sex division and disparity did, however, bestow an unexpected bonus from the research. Primate research has often shown that male and female monkey play tends to appear in different patterns. Males typically play rough-and-tumble, whereas females prefer less contact with a more alternating role of approach and withdrawal, as in the game of tag. Since our female therapists chose the predominantly female forms of play, the males could not have learned from them to wrestle and roll and sham-bite. However, in the final playroom sessions, there were not eight monkeys executing the eloquent elegancies of female games, but just the four who were born female. The four free males who were exhibiting a predominance of age-appropriate masculine play had experienced total recovery.

For some years before the realization of rehabilitation of the 6-month isolates and even for some years afterward, the possibility of successful therapy for monkeys isolated for 12 months was considered to be completely inconceivable. The behavioral deficits of the longer incarceration are decidedly more devastating. The self-directed behaviors of the year-old social isolates were severe to the point of self-aggression, self-biting, and even mutilation. The isolate emerged both fearful and forlorn, barely moving from one fixed and forced abnormal position to another. There was no attempt at all to move around and explore, nor did any incipient indication of social behavior appear at any later date.

Nevertheless, the predictions were proved pessimistic and the rehabilitation of 12-month isolates not only equalled but was confirmed to a degree even beyond that of the 6-month isolates. The experiments of Novak (Novak & Harlow, 1975) were continued into overtime in order to check and confirm not only social adaptation to the therapists, but also to other socially raised rhesus. There were, of necessity, further time extensions to see if sex behavior upon maturation successfully joined the developmental sequence of behaviors. Satisfactory sex adds the crowning criterion to that of normal social play in the social adaptation of the monkey isolates, and both were eventually achieved.

The techniques responsible for this therapeutic triumph were threefold, in addition to the procedures used in rehabilitation of the 6-month isolates. The junior therapists were like the 3-month-old therapists, socially raised with surrogates and allowed to play together for periods both in the quad cage and in the primate playroom. They were, however, even younger in relationship to the 12-month isolates than were the 3-month therapists to their patients. They were even less than one-third as old as the incarcerated infants, 4 months and 13 months at the time of mutual meeting.

There were two preliminary procedures to prepare the isolation-ravaged rhesus (Fig. 124) for companionship. First, the adaptation to changes of environment was even more gradual. The four isolates were in another quad cage in the same room. The isolate subjects could see each other, but an outer wall of solid masonite obscured the therapists from their view. After 2 weeks, this outer covering was replaced with mesh panels with just the lower half covered with 19 inches of masonite. Above the masonite, the recovery animals could contemplate their therapist neighbors if they deigned to look out. To give them encouragement in this direction, the water bottle spout entered the mesh of the cage above the masonite barrier. The isolates were human enough, if not monkey enough, to fall for the foible.

As the monkeys could choose whether or not to climb the walls and whether or not and how frequently they would look at the outer world,

Fig. 124. Twelve-month isolation ravaged rhesus.

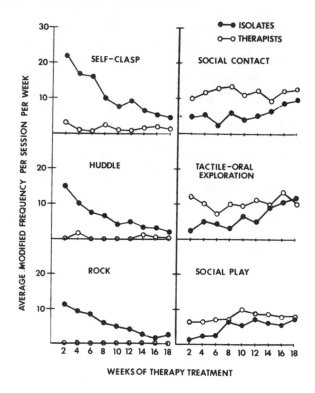

Fig. 125. Disappearance of the undesirable and development of social behaviors during therapy of 12-month isolates.

Novak has termed this technique self-paced therapy, vaguely reminiscent of the self-paced adaptation of the monkeys to the test situation in the learning experiments on the effects of adverse and enriched early environments. The isolates cooperated and paced themselves well, not only climbing the mesh wall to look out and watch the therapists, but watching them most often during the play period together once a day. In this way, without any worry or trauma, the recuperating monkeys, by proxy, first perceived pleasant play.

Self-directed abnormal behaviors suddenly received some respite. It is difficult to clasp oneself while climbing a wall and to suck one's toe while drinking from a dripping spout, and it is just plain boring to ball up in a huddle when one can watch cute young therapists at play.

The isolates were first allowed to be with each other, first in pairs, then all together, 6 weeks after leaving the isolation cages—again a much more gradual approach to a new stage of social rehabilitation. Two weeks later, the slow adaptation to the therapists started. By this time, the isolates were 14 months of age and the therapists only 4 months, the equivalents of approximately 4½ and 1¼ years of human life. A very little child shall show them.

How gradual the adaptation process could be, and necessarily needed to be, is indicated by the approach of one of the isolates to one of the high platforms in the playroom. The recuperating rhesus climbed the ladder and reached out one hand to touch several parts of the platform, then backed bashfully down the ladder. This routine the rhesus repeated about 40 times over a number of test sessions before braving the shelf itself.

Figure 125 presents the complete plotting of the disappearance of the undesirable, infantile, maladaptive behaviors while the equally gradual build-up of positive play and other social graces is occurring. When isolation has continued for one whole year, a long time has passed if this is in human childhood and a year is four times as long for the monkey infant. It is not surprising, then, that three different techniques are required and an equally long time—over a year totally, of therapy, and a tremendous amount of patience to accomplish effective therapy.*

*For further experimental data and details the reader is referred to: Harlow, H. F., & Suomi, S. J. (1971b). Social Recovery by Isolation-Reared Monkeys. *Proceedings of the National Academy of Sciences*, **68**, 1534–1538.

EPILOGUE

FROM THOUGHT TO THERAPY

A basic maxim of scientific investigation is that significant research directed toward providing an answer for a particular question will inevitably generate a host of new problems awaiting resolution. Rarely is a scientific inquiry germinated and subsequently resolved in a vacuum. The endless effort required to solve any major problem frequently leads to other channels of thought and the creation of new areas of interest—often by chance or almost chance associations.

Multiple illustrative cases substantiating this point have evolved from research carried out over the years at the University of Wisconsin Primate Laboratory. We have never completely forsaken any major research goal once we pursued it, and we are still searching for the end of each and every rainbow—even though we have already found our fair share of research gold.

During the Primate Laboratory's 40 years of existence, we have maintained an ongoing research program investigating the learning capability of rhesus monkeys. Learning has been the key directing the creation, not the culmination of many of our major research efforts. The first of a series of studies stemming from the earlier learning researches determined the effects of

273

lesions in specific cortical regions, including unilateral and bilateral occipital (Harlow, 1939), frontal (Harlow & Spaet, 1943), and temporal (Harlow, Davis, Settlage, & Meyer (1952) lobes, on learning task performance. Just as the early lesion research developed from learning, later learning research stemmed from the lesion research. To assess lesion effects, we were forced to create more reliable and lucid learning tasks and to develop and standardize them into a battery of tests that covered varied abilities and cortical locations. A natural problem raised then concerned the ontogenetic development (Harlow, 1959) of ability to perform these various tests, for we already knew that some tests were so difficult that they could not be solved by monkeys younger than 3 years, and some were so simple they could be solved by monkeys in the first weeks of life.

To study developing learning abilities in monkeys required a large number of subjects spanning the age range from birth though adolescence, and so we insituted a breeding program and devised means for rearing monkey subjects in the laboratory from birth onward. In order to reduce the incidence of both confounding variables and contagious disease, we separated the babies from their mothers a few hours after birth and raised them in individual cages where they were hand-fed and received human care (Blomquist & Harlow, 1961). The infants were provided with cheesecloth diapers to serve as baby blankets, and we noticed that many of the neonates developed such strong attachments to the cheesecloth blanket that it was hard to tell where the diaper ended and the baby began (Fig. 126). Furthermore, the monkeys became greatly disturbed when the diapers were removed from their cages for essential sanitary services.

Dirty diapers and distressed infants were produced for some years—an adequate time for insightful learning to occur—before the true significance of the diaper was duly recognized. It is a long way from brains to blankets, but this is the strange, mysterious way in which research programs develop. Many creative ideas have suddenly appeared in a flight of fancy, but the surrogate mother concept appeared during the course of a fancy flight. The cloth surrogate mother was literally born, or perhaps we should say baptized, in 1957 in the belly of a Boeing stratocruiser high over Detroit during a Northwest Airlines champagne flight. Whether or not this was an immaculate conception, it certainly was a virginal birth. The senior author turned to look out the window and saw the cloth surrogate mother sitting in the seat beside him with all her bold and barren charms. The author quickly outlined the researches and drafted part of the text and verses which would form the basis of his American Psychological Association presidential address (Harlow, 1958) a year later. The research implications and possibilities seemed to be immediately obvious, and they were subsequently brought to full fruition by three wise men—one of whom was a woman.

Fig. 126. Infant monkey clothed in cheesecloth.

The original theoretical problem to be solved by the surrogate mother researches was to measure the relative strength of bodily contact comfort as opposed to satisfaction of nutritional needs, or activities associated with the breast, as motivational forces eliciting love for mother in rhesus neonates. Actually the primary purpose was to continue to dismantle derived drive theory (Harlow, 1953). The results of the now famous cloth-mother and wire-mother experiments provided total support for contact comfort as the superordinate variable or motive binding infant to mother. As pictures of baby monkeys clinging contentedly to soft surrogates (see Fig. 127) unfolded across tabloid pages throughout the world, the downfall of primary drive reduction

Fig. 127. Infant monkey clinging to cloth surrogate.

as the predominant theory to account for the development of social attachment was assured. The cloth mother became the first female to attain fame so quickly while still retaining her virginal virtues. There is more than merely milk to human kindness.

On the basis of the diaper data, it came as no great surprise to find that monkey infants overwhelmingly preferred nonlactating cloth mothers to lactating wire surrogates. However, during the course of testing infants in novel environments, we discovered an unexpected trait possessed by our cloth surrogates: the capacity to instill a sense of basic security and trust in their infants (Harlow & Zimmerman, 1959). This is the way creative research often

Fig. 128. Infant terrified in absence of cloth surrogate.

arises—sometimes by insight and sometimes by accident. Baby monkeys placed in an unfamiliar playroom devoid of a cloth surrogate, or with a wire surrogate present, typically rolled into tight furry balls, (Fig. 128) screeching in terror.

When the same infants were placed in the same environment in the presence of cloth surrogate mothers, they initially scurried to the surrogates and clung for dear life. After their first fears abated, the monkeys would then venture away from the surrogates and explore the environment, as shown in Fig. 129, but often returned (Fig. 130) to their inanimate mothers for a reassuring clasp or a deep embrace to desensitize fear or alleviate insecurity. This response was predicated upon a psychiatric principle discovered by baby monkeys long before the advent of Watson (1924), Wolpe (1958), or any of the Skinnerians. Basic trust was the achievement of the first of Erikson's (1950) eight human developmental crises, and although basic trust may not be fashioned out of whole cloth, for baby monkeys it apparently can be fashioned from cloth alone.

Fig. 129. Infant security in presence of surrogate mother.

Subsequently, we recognized the obvious truth that no major act of animal behavior is determined by a single variable. To illustrate this axiom, we created surrogates of varying form and function, and they disclosed that many variables other than contact comfort possessed more than measurable effects on infant monkey maternal attachment (Furchner & Harlow, 1969). These findings led naturally to a series of studies designed to measure all possible variables, regardless of importance, relating to the maternal efficiency of our man-made mothers. The researches disclosed a number of variables secondary in importance to contact comfort. With contact comfort held constant by constructing lactating and nonlactating terry-cloth surrogates, it was possible to demonstrate that nursing, or activities associated with the breast, was a significant variable during the first 90 days of life. Thus, by this ingenious research we learned what had been totally obvious to everyone else, except psychologists, for centuries. Furthermore, rocking surrogates and rocking cribs were preferred to nonrocking surrogates and cribs for about 160 days. Body surfaces other than wire or cloth were also investigated, with predictable

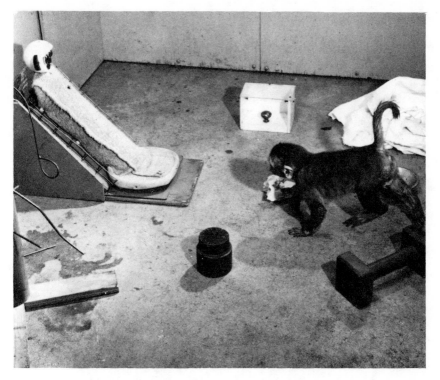

Fig. 130. The infant often returned to rub in some more security.

results. Satins and silks might be adult symbols of prestige, but they do not warm the infant heart as does terry cloth.

Infant rhesus monkeys preferred a warm wire surrogate to a cool cloth surrogate for the first 15 days of life, illustrating the limited temporal span of some variables and confirming the well-known "hot mama" or "warm woman" hypothesis. Warmth was the only variable to lend even transient preference to the wire surrogate. However, the most striking maternal temperature data were recently obtained by Suomi (Harlow & Suomi, 1970), who constructed a cold cloth surrogate with ice water in her veins. Neonatal monkeys tentatively attached to this cold cloth figure, but then retreated to a far corner of the cage (Fig. 131) and remained aloof from mother forever. There is only one social affliction worse than an ice-cold wife, and that is an ice-cold mother.

Finally, we compared the efficiency of our man-made mothers with their natural counterparts, and we are convinced that real motherhood is superior and that it is here to stay. The cloth mother may serve milk, but not in the

Fig. 131. Baby rejecting cold surrogate mother.

cozy continuous containers provided by the real mother. The real mother eliminates nonnutritional sucking by her infant, whereas no surrogate mother, regardless of Skinnerian schedule, can inhibit nonnutritional sucking. The real monkey mother trains her infant to be a placer, rather than a spreader, of feces (Hediger, 1955). The real mother trains her infant to comprehend the gestural and vocal communications of other monkeys (Miller, Murphy, & Mirsky, 1959), while language learning is beyond surrogate love. The real mother is dynamic and responds to the infant's needs and behavior, but the surrogate can only passively accept. Subsequently the mother plays an active role in separating the infant from her body, which results, probably inadvertently, in the exploration of the surrounding animate and inanimate environment. Finally, and of most importance for future peer adjustment, the real mother is far more efficient than the cloth surrogate in the regulation of early infant play, the primary activity leading to effective agemate love.

We might have remained imprinted on surrogate mothers forever had it not been for a comment made to the senior author independently by an eminent psychologist and an eminent psychiatrist within a single month. Both said, "You know, Harry, you are going to go down in the history of psychology as the father of the cloth mother!" This was too much! In a desperate effort to escape this fate, we branched out into new areas of research which were subsequently to fall into two broad, disparate areas: the nature of normal and natural love in rhesus monkeys, and the induction of psychopathology.

I enlisted the aid of my wife, and we fell in love, or at least in love with love, in all its multifaceted forms. Normal and natural love in rhesus monkeys develops through the sequencing and interaction of five major love systems: maternal love; infant love, or love of the infant for the mother; peer love, which other psychologists and psychoanalysts will someday discover; heterosexual love; and paternal love.

Maternal love has always been obvious, and even Freud was fully aware of it. We have already described its social functions. An extremely important basic function is the management of infant play so that infant monkeys play together effectively instead of in a disorganized manner. Maternal love serves as an important antecedent to the development of peer or agemate love.

The variables underlying the love of the infant for the mother have already been described in the surrogate researches. It is our opinion that a more important love system, in fact, the most important from the view of the whole life-span, is agemate love, which develops first through curiosity and exploration and later through multiple forms of play. Peer interactions enhance the formation of affection for associates, the development of basic social roles, the inhibition of aggression, and maturation of basic sexuality. We believe heterosexual behavior in primates is another love system, evolving from peer love very much as peer love evolves from maternal love.

Heterosexual love was not discovered by Freud. Freud became lost in the libido even before he experienced it, and he never fully learned about love. Heterosexual love differs in form and function in various animal families. Beach (1969) eventually discovered love in beagles, but so had the beagles. Some female beagles have warm and wondrous love affairs despite the fact that they are basically bitches. Heterosexual love in rats and people is planned in different fashions. If you are a rat, your sex life may be endocrinologically determined, and you will do very well. However, if you are a primate—monkey, ape, or man—and your heterosexual life is primarily gonadally determined, you face a grim and grave future, and the sooner the grave, the better. Sex without antecedent and concurrent love is disturbed and disordered (Harlow, 1965).

After resolving the nurture and nature of maternal, infant, peer, and heterosexual love, the only thing that remained was paternal love. Having analyzed monkey love as far as we could with our existing facilities, I realized that we had no love with father, and I dejectedly proclaimed to my wife that, although paternal love in feral baboons and monkeys had been described, this love system could not be analyzed and resolved under laboratory restraints. A month later Margaret Harlow brought me the experimental design for paternal laboratory love and a plan for the necessary housing facilities. After the relatively simple task of rebuilding the attic over our laboratory had been achieved, the analysis of paternal love was on its way.

The nuclear family apparatus is a redesigned, redefined, replanned, and magnified playpen apparatus where four pairs of male and female macaques live with their offspring in a condition of blissful monogamy. In the nuclear family apparatus, each and every male has physical access to his own female and communicative access to all others. It is obvious from time to time that some males and females would like to have physical access to their neighbors' mates, but their courting must be limited to calls and lip-smacking and visual fixation. Fortunately, they accept their frustration with minimal effects on their mates.

Most important of all, each and every infant has access to every male, and, perhaps because of the cunning and curiosity of all the infants, most nuclear fathers responded socially to most infants. Finally, the apparatus provides unrivaled opportunities to study sibling interactions and friendship formation in infants of similar and disparate ages.

Creation of the nuclear family has provided us with a body of basic information concerning paternal love. The nuclear fathers do not allow mothers, their mates, and neighbors, to abuse or abandon infants, and the fathers serve as a cohesive force guarding the group against predators—primarily experimenters. In addition, the fathers, through some developmental mechanism which we do not yet understand, show affection in varying degrees to all infants. Many fathers engage in reciprocal play with the infants at a level far surpassing that of the mothers, and the fathers ignore aggression from the infants and juveniles, including pinching, biting, and tail- and ear-pulling—behaviors the fathers would never accept from adolescents and adults of either sex.

Preadolescent monkey males, unlike females, exhibit limited interest in all new infants except their siblings until the babies can play. The males largely ignore them, while the female preadolescents continually struggle to make contact with the new babies. The precursors of paternal behavior are present, however, for the older male infants and juveniles cradle, carry, and protect young infants that venture in their path. The watchful eyes of the adults and their ready threats may abet the gentle behavior of the older infants and possibly begin the inculcation of protection of all young. We have still much more to learn about the variables in the development of paternal behavior.

The advent of the second and third infants in the families has disclosed interesting aspects of maternal love and sibling interaction. We had long presumed that the appearance of a second monkey gift from heaven would exaggerate the mother-infant separative mechanisms long in progress, and that neonatal fairy fingers playing upon the maternal heartstrings would rapidly dissipate the love for the older infant. True to prediction, the immediate reaction of the newly delivered mother to her older infant was negative. She threatened body contact, prevented nipple contact, and cradled the new infant continuously. But every mother eventually reversed this policy toward the older infant. The only individual difference was the interval between the new birth and contact with the older infant, which ranged from 8 hours to a matter of days. Most displaced infants or juveniles spent a night or two without maternal contact, often with their fathers, but one managed to achieve contact with mother the very first night and every night thereafter by persistent approach, cooing, and squealing until her mother made room for her too. Although she had a good relationship with her father, she made no attempt to substitute him.

Much to our surprise, the displaced infants did not overtly exhibit punitive signs of jealousy toward the newcomers, probably because of fear of the mother, although one male juvenile did engage in teasing his little sister at every opportunity when mother was not looking. All displaced infants showed disturbance in this situation of denial and despair, of suspicion and separation, and the older infants would spend hours trying to achieve contact comfort, real or symbolic, from the body of the mother—both awake and asleep. Indeed, initial contact was usually made when the mother was sleepy and had reduced her vigilance. In desperation, when this failed, some would enter adjacent living chambers and make overtures to other mothers, who generally accepted their presence but denied them bodily contact. Alternatively, proximity and contact with their fathers were sought when mothers were not available.

In spite of the fact that the nuclear families provided a wealth of new data on the affectional systems, the most striking psychological contribution of the nuclear family has not been to love but to learning.

For a number of years we had assiduously studied the effects of early environment upon later learning capability, and to achieve this we had always used groups of normal monkeys and groups of socially isolated monkeys. We knew that total social isolation damaged or destroyed the social-sexual capabilities of monkeys, but it did not depress learning ability. Our socially deprived monkeys were reared under conditions of 6, 9, or 12 months of total social isolation, a condition of deprivation or privation so severe that no one will ever impose it upon human children.

Our "normal" monkeys had been reared in partial social isolation. We had recognized the fact that partial social isolation would hardly qualify as a haven or Heaven, but because of limited facilities this is the manner in which we had always reared our normal monkeys. For decades our normal monkeys have achieved learning performances better than those achieved in any other laboratory, owing, no doubt, to the unusual care we took in adapting them to the test situation.

Finally, S. D. Singh (1969), who had had extensive test experience on the Wisconsin General Test Apparatus (WGTA) in the United States and in India, reported that feral animals (reared in forests or in temples) were not intellectually different from each other and were not superior to our monkeys reared in partial social isolation. Furthermore, Singh's test battery was adapted from our own, utilizing discrimination problems which rhesus monkeys are able to solve at 6 months of age, delayed-response tests at 10 months of age, learning-set tasks mastered at 12 months of age, and, later, complex oddity-learning set tasks which are not efficiently solved by monkeys until 36 months of age. Singh's data gave every indication that partial isolation cages were just as stimulating to intellectual development as were temples and forests.

We had assumed that "enriched" environments were in no way superior to the deprived environments in stimulation and development of the intellectual processes. To demonsgrate this, we compared the performance of monkeys reared from birth in the nuclear family apparatus with that of totally socially isolated monkeys and our normal monkeys. Just as predicted, the enormously socially enriched monkeys reared in interacting family groups did no better than deprived monkeys or control monkeys on discrimination tasks, delayed-response tasks, and complex learning-set tasks. My world of happy intellectual isolation was jolted, however, when the socially enriched preadolescents and adolescents, as contrasted with the socially isolated adolescents and controls, proved to be superior at the .001 significance level on our most complex problem-oddity-learning set. Had there been a progressive separation in performance between enriched and deprived monkeys as they traversed through tests of increasing complexity, we would gladly have conceded a difference, but the difference appeared only when the most complicated learning test was administered.

One can only conclude that this enriched early environment, at least, enables monkeys adequately adapted and trained to reach more lofty intellectual performance levels than those attained by deprived monkeys. The basis for the performance difference, however, is by no means established. Superiority could stem from nonintellectual factors as readily as from intellectual differences. The nuclear family animals give every evidence of being the most self-confident, self-assured, fearless animals we have ever tested. They are more relaxed in the test situation than other subjects and could well be more persistent, thus persevering after "normal" subjects give up. This difference would not be apparent on unchallenging tasks, but when the problems become very difficult, the personality factors could operate to produce performance differences. Unfortunately, it is as difficult to test as is the hypothesis that middle-class children excel intellectually over lower-class children because of their environmental advantages.

For many years behavioral scientists attempted to produce psychopathological behavior syndromes in nonhuman subjects by experimental manipulation, but their successes were at best limited. H. F. Harlow (1971) hypothesized that maternal rejection might provide the critical contribution to this area, and so a family of surrogate mothers was designed to impart fear and insecurity to infant monkeys. Four different forms of evil artificial mothers emerged, and although all were designed to repel clinging infants, each had its own unique means by which to achieve this end. One surrogate blasted its babies with compressed air, another tried to shake the infant off its chest, a third possessed an embedded catapult which periodically sent the infant flying, while the fourth carried concealed brass spikes beneath her ventral surface which would emerge upon schedule or demand.

These surrogates produced temporary emotional disturbance in the infants, but little else. When displaced from their artificial mothers, the infants would cry, but they would return to the mother as soon as she returned to normal. In retrospect, it should have been obvious—to what else can a frightened, contact-seeking infant cling? The only individuals to suffer prolonged distress from these experimental efforts were the experimenters. Later we discovered the existence of far more sadistic monkey mothers—real ones.

In the midst of these ill-destined efforts, we discovered that we already had dozens of abnormal monkeys produced without any special effort. It became apparent to us that something was awry after the animals we tested in the study of the development of learning had completed their experimental chores and were physically mature enough to transfer to the breeding colony. We had every expectation that these healthy, well-developed animals would give us disease-free infants to supply our experimental needs. The animals were paired appropriately and placed in large cages. Weeks passed, then months passed, and we never saw any copulation, and there certainly were no offspring. When summer came, we hoped to change their behavior by assigning some of them to group living on an unoccupied monkey island in the Madison zoo.

The group psychotherapy had some effect. The aggression that erupted when the monkeys were initially transferred gradually disappeared. Animals began to form social groups and to groom each other. But no heterosexual behavior was observed and there were no pregnancies. In the belief that a highly experienced male from the breeding colony could conquer the females, we released onto the island one of our most capable males. He met all challengers with ease and immediately became leader of the island. But still no females became pregnant. We knew then that we had raised a colony of monkeys abnormal in their sex behavior. This was the beginning of a systematic effort to study sexual and social development of monkeys raised under varying environmental conditions.

Fig. 132. Catatonic posture of partial isolate.

One rearing condition we had already established—the raising of infants from birth onward in bare wire cages without companions. Subsequently, we termed this "partial social isolation." The subjects were not only devoid of heterosexual behavior at maturity but showed exaggerated oral activities, self-clutching, and rocking movements early in life, then apathy and indifference to external stimulation subsequently. Individualized stereotyped activities involving repetitive movements characterized many subjects and extremely bizarre behavior appeared in some. An animal might sit in the front of its cage staring aimlessly into space. Occasionally one arm would slowly rise as if it were not connected to the body, and wrist and fingers would contract tightly (Fig. 132)—a pattern amazingly similar to the waxy flexibility characteristic of some human catatonic schizophrenics. The monkey would then look at the arm, jump away in fear, and subsequently attack the offending object. Unfortunately, we know nothing about the forces that cause one isolated monkey to drift into inactivity and another to demonstrate bizarre repetitive behavior patterns.

To understand why partial social isolation, which seems to be a relatively benign condition, is actually so devastating socially, one need merely consider the effect that rearing in partial or total social isolation from birth onward has upon the development of the five love systems (Harlow, 1971). The monkey reared in partial social isolation knows no maternal love and therefore cannot love its mother. Furthermore, living alone in a cage it cannot develop agemate or peer affection, which comes, for the normal monkey, through physical interaction with other young monkeys. Sex in a bare wire cage is either nonexistent or at best limited and lonely. Thus it is not surprising that partial social isolation produces profound behavioral abnormalities in monkeys. By chance, we had discovered what had been sought for years by design.

If denial of physical access to other monkeys produces such psychopathology, it would seem likely that denial of visual, as well as physical, access to other monkeys would produce even more serious deficiencies. Subsequent research found this to be the case. When William A. Mason came to Wisconsin in 1954, we designed total isolation chambers, and the first research describing the effects of long-term social isolation in these chambers was published by Mason and Sponholz (1963). Subsequently, an improved total social isolation apparatus was created with true cunning and connivance by Rowland (1964), and this became and remains our standard total social isolation chamber. Rowland's apparatus was designed so that monkeys could be raised from birth onward without seeing any other animal except the experimenter's hands and arms which assisted the neonate up a feeding ramp during the first 15 days of life. Subsequently, the experimenter could easily observe the infant through a one-way vision window, while the infant monkey saw no animal of any kind. Moreover, the apparatus was designed so that the learning ability of the isolate-reared monkey could be measured by remote control, and this was successfully achieved in Rowland's original study.

The total social isolation apparatus enabled us to quantify the socially destructive effects of isolation from birth onward. Monkeys reared in total social isolation for 90 days were enormously disturbed when admitted to the great wide world of wonder, and two of them actually died of self-induced anorexia before we recognized the syndrome and instituted forced-feeding. However, all surviving monkeys rapidly made a complete social adjustment to agemates, so that behavior from one postisolation month onward was normal for all measurable purposes.

In contrast, monkeys subjected to 6 months of total social isolation from birth and then allowed to interact with agemates were very adversely affected for the rest of their lives. They spent their time primarily engrossed in autistic-like self-clasping, self-mouthing, and rocking and huddling. The isolates never interacted successfully with normal peers over an 8-month period, although pairs of isolate monkeys did show limited recovery in terms of

Fig. 133. Isolate infant frozen with fear.

exploration and even play with each other. These monkeys were then maintained in partial social isolation for approximately 3 years, and, when subsequently tested, their behavioral repertoire appeared to have deteriorated rather than improved. Their social efforts were plaintive and their sexual efforts pitiful. Practically the only social behaviors that seemed to have matured were fear and aggression, and the animals showed these inappropriately and often explosively. Six-months isolates aggressed against infants—an act no normal monkey would consider—but before, during, and after aggressive acts, they were frozen in fear even though the infants they faced were only half their size. In addition, several isolate monkeys attempted a single suicidal sortie against very large adult males—an act no normally socialized animal would be foolish enough to try (see Fig. 133).

We also discovered that 12 months of total social isolation from birth had even more drastic effects than 6 months on behavior in the playroom. Exploration and even simple play were nonexistent. Torn by fear and anxiety, aggression was obliterated in these monkeys, and even the simple pleasure of onanism was curtailed. They sat huddled alone in the corners or against the walls of the room. The actual experiment was stopped after 10 weeks, since the control animals were literally tearing up the 12-month isolates, and the isolates made no effort to protect themselves. These animals were maintained for many years and never demonstrated any vestige of virginal social ability, even for a very long time afterward. They were tested by Robert E. Miller some 3 years later for their ability to receive and express social communication. They were a perfect control group, totally devoid of all social signaling.

A considerable number of our isolate-reared females were eventually impregnated by patient and competent feral males. When adequate animal assistance failed, we resorted to an apparatus that restrains, positions, and supports the female during copulation. Very soon we discovered that we had created a new animal—the monkey motherless mother. These monkey mothers that had never experienced love of any kind were devoid of love for their infants, a lack of feeling unfortunately shared by all too many human counterparts (Helfer & Kempe, 1968). Most of the monkey motherless mothers ignored their infants, (Fig. 134), but other motherless mothers abused their babies by crushing the infant's face to the floor, chewing off the infant's feet and fingers, and in one case by putting the infant's head in her mouth and crushing it like an eggshell. Not even in our most devious dreams could we have designed a surrogate as evil as these real monkey mothers.

The door to discovery of behavioral deficits produced by isolation rearing was opened largely by chance. In contrast, the initial enlightenment of procedures which resulted in another form of monkey psychopathology, that of depressive behavior, was unlocked by love. During the course of study of mother and infant affection, two experiments (Seay et al., 1962; Seay & Harlow, 1965) were conducted in which infants were reared with mothers and peers, then separated from their mothers for a period of several weeks. It was found that the maternal separations precipitated severe reactions among the infants. During the period of separation, the young monkeys ceased their peer play activity and became withdrawn and inactive. When reunited with their mothers, they spent more time engaging in mother-directed activity than they had spent immediately prior to separation, amply demonstrating the overwhelming strength of the mother-infant attachment bond.

A closer examination of these data indicated that more than love had been disclosed by the experiments. During the maternal separation the infants had initially expressed protest, characterized by increased activity and vocalization, but soon withdrew and became inactive. Normal social interactions among the infants declined or disappeared, as was the case for play, the most complex social behavior the infants possessed. These behavioral abnormalities vanished when the infants were reunited with their mothers.

Fig. 134. Motherless mother ignoring her crying baby.

Several years earlier Spitz (1946) and Bowlby (1960) had witnessed surprisingly similar reactions among human infants who had been separated from their mothers via hospitalization. Spitz termed the reaction "anaclitic depression." Bowlby delineated two stages of the reaction during the period of separation, which ranged from a few days to a few months: initial protest, characterized by agitation and crying, and despair, characterized by withdrawal from the world of both animate and inanimate objects. When Spitz's children were reunited with their mothers, recovery was immediate and spontaneous, but Bowlby observed a "detachment" among some of his infant patients upon maternal reunion, a phenomenon he now feels may not represent a universal aspect of reaction to maternal separation. At any rate, these data suggested a close parallel between human and monkey infants in terms of reaction to maternal separation: Anaclitic depression resulted in both cases.

Mother-infant separation experiments were subsequently carried out at several other primate laboratories, and the findings from study to study were amazingly consistent (Kaufman & Rosenblum, 1967; Hinde et al., 1966). Almost immediately following separation, infants exhibited initial protest, characterized by increasing activity and vocalization. Shortly thereafter, most subjects entered into a depressive withdrawal, even though the form and duration of the despair stage varied among the monkeys in the various researches and differing experimental conditions. Upon reunion, infant-mother interactions rapidly became essentially normal and there was little evidence of maternal detachment.

Several years later, the accumulation of a vast body of normative information and a desire to investigate new and challenging problems led us to shift our major interests to the study of depression in monkeys. Leaving love in search for psychiatry is not that major a transition. The study of normal behavior furnishes all needs and norms necessary for the analysis of monkey madness and laboratory lunacy. At this time, a reconsideration of the mother-infant separation studies indicated a point of departure for the production of depression in nonhuman primates. Our earlier studies had duplicated, in the laboratory monkeys, both the precipitating situation and the subsequent syndrome described as anaclitic depression for human infants. However, it was obvious to us that mother-infant separation had both theoretical and practical limitations as a standard procedure for large-scale investigation of depression in monkeys, and to achieve significant advances in this area it would be necessary to transcend the mother-infant separation model.

Our first effort in this direction was initiated by Suomi (Suomi et al., 1970), who reared infant monkeys with each other rather than with mothers. When these infants were separated from their playmates at 3 months of age, they exhibited a protest-despair reaction to separation virtually identical in form to that resulting from maternal denial in both human and monkey infants. Unlike the mother-infant separation studies, the infant-infant separation technique was expanded so that the young monkey peers were separated from each other not once but many times—4 days for each of 20 experimental weeks spread over a 6-month period. During every separation period, the infants exhibited a severe protest-despair reaction, and each time they were reunited their activity primarily consisted of mutual clinging. This pattern did not change significantly from the beginning to the end of the multiple separation periods, even after 20 separations.

An unanticipated and fascinating discovery was the finding that the multiple separations produced a severe maturational arrest in the monkeys.

Their behaviors following the separations were as infantile at age 9 months as they were prior to the first separation. Neonatal behaviors of nonnutritional orality and self-clasp persisted throughout the 6-month separation period, but the complex infant play activities, which normally mature from 90 to 180 days, had not appeared by the age of 9 months. This finding was in total contrast to the fascinating progression of social development traditionally reported in normal monkey infants. It was as if Suomi had stopped the monkeys' biological calendars.

The results of Suomi's study indicated that depressive reactions could be precipitated in monkey subjects by procedures other than that involving separation from the mother. No longer bound by the restraints of the mother-infant separation model, we could now seriously consider production and study of depressions other than anaclitic in monkey subjects.

A radically different approach to the production of depressive behavior in monkeys, one that did not involve any social attachments, was made possible by a vertical chamber apparatus created by H. F. Harlow. This apparatus is a stainless steel chamber with sides that slope downward to a wire-mesh platform above a rounded steel bottom. Depression in humans has been characterized as a state of "helplessness and hopelessness, sunken in a well of despair" (Schmale, 1971), and the chambers were designed to reproduce such a well for monkey subjects. Although the confined monkeys are free to move about in three dimensions within the chamber, and although they eat and drink normally and maintain proper weight, within a few days they typically assume a huddled, immobilized posture in a corner of the apparatus.

Suomi (1973) then tested 90-day-old monkeys antecedently subjected to 45 days of chamber confinement and compared their subsequent activity in both social and nonsocial situations with two groups of equal-aged monkeys, one group peer-reared and the other reared as partial isolates. Extensive testing was conducted for 9 months, and throughout this period the chambered subjects consistently exhibited highly elevated levels of self-clasping and huddling, low levels of locomotion and exploration, and nonexistent social activity. These behaviors were in sharp contrast to those of both control groups (Fig. 135). Clearly, chamber confinement of relatively short duration was enormously effective for producing profound and prolonged depression in young monkey subjects.

Suomi then measured the effects of combined chamber confinement and peer separation in two studies utilizing monkey subjects under a year of age and found that chamber confinement intensified depressive separation-produced effects in monkeys with extensive prior social experience. We have long believed that a phenomenon as complex as depression cannot possibly be mediated by a single variable, and these researches indicated that the depths of our infants' depressions were dependent upon a number of

PLAYROOM BEHAVIORS OF RHESUS
MONKEYS 9-11 MONTHS OF AGE
REARED UNDER DIFFERENT CONDITIONS

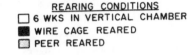

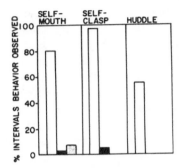
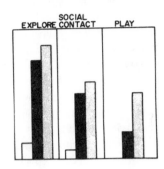

Fig. 135. Behavioral effects of chamber confinement; playroom behaviors of rhesus monkeys 9-11 months of age reared under different conditions.

factors, including duration of separation and/or confinement, age at which depression was produced, and prior social history of the subjects.

The addition of a psychiatrist, Dr. William McKinney, to our depression project, brought both clinical insight and psychiatric respectability to the research endeavors, and his presence has been welcome and fruitful. He has already taken a leading role in the experimental induction of depression in older-aged monkeys, whereby the diagnosis of anaclitic depression is excluded. In addition to investigating the behavioral aspects of depression, he has directed the initiation of researches involving analyses of biochemical variables, including the catecholamines (McKinney, Eising, Moran, Suomi, & Harlow, 1971). Three years ago, the idea of using monkeys to unravel the behavioral and biochemical intricacies of an affliction suffered in some form, and at some time, by virtually every human being and fully understood by virtually no human mind, seemed to be little more than a desperate dream or humble hope. Today we are finally and firmly on the road to success.

Our research endeavors in the field of depression have resulted in a sizable increase in our population of emotionally disturbed monkeys, and we are now initiating researches designed to rehabilitate our "patients" to a state of social normality. We plan long-term researches utilizing all possible types of therapeutic agents, including various antidepressant drugs and even such techniques as electroconvulsive therapy (ECT). Actually, our primary interest lies in the alleviation of depression by psychotherapeutic techniques since many of the monkey depressions were induced through social manipulations. Thus, social approaches to therapy have been our principal concern. Recently we have had several rewarding experiences which have provided valuable information relative to the stages of depression formation and alleviation.

While we were inducing depression in monkeys, we were also attempting to rehabilitate several of our total social isolation-reared subjects, an effort previously initiated by many investigators with little or no success (Sackett, 1968). As stated earlier, isolates exposed to socially normal agemates were the recipients of severe aggression with little therapy, and any subsequent social improvement was limited at best. However, we did discover that isolate-reared subjects showed some social improvement if they were able to achieve contact acceptability with various social agents. In particular, the motherless mothers whose infants survived in spite of evil maternal efforts eventually submitted to their babies' persistent attempts to achieve and maintain maternal contact, and to our great surprise these females usually exhibited adequate maternal behavior toward subsequent offspring. Also, isolates exposed to heated surrogates, soon after emergence from confinement, eventually learned to contact the nonthreatening surrogates, leading to significant decreases in self-directed disturbance activity. When these animals were subsequently paired with each other, they exhibited the rudiments of basic social-interaction patterns (Suomi, 1973).

These findings convinced us that significant rehabilitation of isolate-reared subjects via social exposure was feasible and that the crucial variable lay in the nature of the social stimulation utilized. Each of the authors is convinced that he (she) created the plan and procedure for monkey rehabilitation, and this is probably true since all were thoroughly familiar with the essential maturational data underlying a feasible therapy program. It seemed that an effective monkey "therapist" might be one who would instill contact acceptability in the isolate monkeys without providing a threat of aggression, and could subsequently or simultaneously provide an appropriate medium for the development of an increasingly sophisticated social repertoire. Our knowledge of monkey social development led us to select socially normal animals 3 to 4 months old as therapists for the isolates. At this age, normal monkeys are too young to show aggression, they still provide stable contact clinging responses, and they are on the verge of gradually expanding their basic social interactions into fully developed play (Harlow & Harlow, 1965).

We therefore took 6-month-old animals who had been socially isolated from birth and housed them individually in compartments of a "quad" cage (Suomi & Harlow, 1969) adjacent to therapists 3 months their juniors. The quad cage, designed by Suomi, is an extremely versatile social testing unit which can simultaneously or successively serve as a living and testing area. Selection of interior panels dictates roommate assignment, while removal of the panels permits partner interaction in home territory. In this study (Harlow & Suomi, 1971) we permitted interactions between isolates and therapists 2 hours per day, both within the quad cages and in a social playroom. As therapy progressed, interaction time in the quad cages was decreased and time in the playroom increased.

The isolates' initial response to their interaction opportunities was to retreat to a corner and rock and huddle, and the therapists' initial response was to follow and cling to the isolates. Soon the isolates were clinging back, and it became only a matter of weeks until isolates and therapists were playing enthusiastically with each other. During this period, most of the isolates' previously abnormal behaviors gradually disappeared, and after 6 months recovery was essentially complete.

An interesting sidelight or fringe benefit from the above findings concerned observed sex differences. Not entirely by design, all of the isolates in the above experiment were males, while all of the therapists were females. We have known that under normal rearing conditions males develop a rougher and more contact-oriented form of play behavior than females, and these differences are initially expressed prior to 6 months of age. Our rehabilitated males had spent their first 6 months of life in total social isolation and thereafter were exposed only to the female therapists and to each other. The psychotic monkeys had no social model for the development or creation of masculine play. Nevertheless, their play, when it emerged, was clearly masculine in form, adding to our data long cumulating that sex-typing of play in monkeys is governed not by imitation but by genetics (Harlow & Rosenblum, 1971). Culture makes clothes but God gives gonads.

Thus we have traveled from thought to therapy by a route neither straight nor narrow. There have been obstacles and detours, but we have found throughout the years that these are to be cherished, not chastised, as blessings in disguise. We began with learning which led to lesions and later to love. Our first love was a soft and simple surrogate. Now it is a sophisticated simian society, whose study has led us back to learning. There have been other grand and great circles. In our study of psychopathology, we began as sadists trying to produce abnormality. Today we are psychiatrists trying to achieve normality and equanimity. Tomorrow there will be new problems, new hopes, and new horizons. Since knowledge is itself forever changing, the search for knowledge never ends.

BIBLIOGRAPHY

Beach, F. A. (1945). Current concepts of play in animals. *American Naturalist,* *79,* 523–541.

Beach, F. A. (1969). Locks and beagles. *American Psychologist, 24,* 921–49.

Berkowitz, L. (1971). The contagion of violence: An S-R mediational analysis of some effects of observed aggression. In W. Arnold & M. Page (Eds.), *Nebraska symposium on motivation,* 1970. Lincoln, Nebraska: University of Nebraska Press.

Berkowitz, L. (1973). Words and symbols as stimuli to aggressive responses. In J. F. Knutsen (Ed.), *Control of aggression: Implications from basic research.* Chicago: Aldine Atherton.

Bierens de Haan, J. A., & Frima, M. J. (1930). Versuche uber den Farbensinn der Lemuren. Z. f. vergl. *Physiol., 12,* 603–631.

Biller, H. B. (1970). Father absence and the personality development of the male child. *Developmental Psychology, 2,* 181–201.

Biller, H. B. (1974). Paternal deprivation: Family, school, sexuality, and society. Lexington, Mass.: Heath.

Bingham, H. C. (1932). *Gorillas in a native habitat.* Washington, D.C.: Carnegie Institute, No. 426, August.

Birch, H. C. (1945). The relation of previous experience to insightful problem solving. *Journal of Comparative Psychology*, **39**, 15-22.

Blomquist, A. J., & Harlow, H. F. (1961). The infant rhesus monkey program at the University of Wisconsin Primate Laboratory. *Proceedings of the Animal Care Panel*, **11**, 57-64.

Blurton Jones, N. G. (1967). An ethological study of some aspects of social behavior of children in nursery school. In D. Morris (Ed.), *Primate ethology*. Chicago: Aldine.

Bowlby, J. (1960). Grief and mourning in infancy and early childhood. *Psychoanalytic Study of the Child*, **15**, 9-52.

Bowlby, J. (1969). *Attachment and loss: Attachment* (Vol. 1). New York: Basic Books.

Boycott, B. B., & Young, J. Z. (1950). The comparative study of learning. In Physiological mechanisms in animal behavior. *Symp. Soc. Exp. Biol.*, No. 4. Cambridge, Eng.: Cambridge University Press.

Bromer, J. (1940). *A genetic and physiological investigation of concept behavior in primates*. Unpublished doctoral dissertation, University of Wisconsin.

Brown, A. L. (1970). Subject and experimental variables in the oddity learning of normal and retarded children. *American Journal of Mental Deficiency*, **75**, 142-151.

Burlingham, D., & Freud, A. (1942). *Young children in wartime*. London: Allen & Unwin.

Butler, R. A. (1953). Discrimination learning by rhesus monkeys to visual-exploration motivation. *Journal of Comparative & Physiological Psychology*, **46**, 95-98.

Buxton, C. E. (1940). Latent learning and the goal gradient hypothesis. *Contributions to Psychological Theory*, **2**, No. 2.

Cannon, W. B. (1929). *Bodily changes in pain, hunger, fear, and rage*. New York: D. Appleton.

Cannon, W. B. (1932). *The wisdom of the body*. New York: Norton.

Carmichael, L. (1934). An experimental study in the prenatal guinea pig of the origin and development of reflexes and patterns of behavior in relation to the stimulation of specific receptor areas during the period of active fetal life. *Genetic Psychology Monographs*, **16**, 337-491.

Carpenter, C. R. (1934). A field study of the behavior and social relations of howling monkeys. *Comparative Psychology Monographs*, **10**, 1-168.

Carpenter, C. R. (1942a). Sexual behavior of free-ranging rhesus monkeys. I. Specimens, procedures and behavioral characteristics of estrus. *Journal of Comparative Psychology*, **33**, 113-142.

Carpenter, C. R. (1942b). Sexual behavior of free-ranging rhesus monkeys. II. Periodicity of estrus, homosexual, autoerotic and nonconforming behavior. *Journal of Comparative Psychology*, **33**, 143-162.

Clark, W. E. LeGros (1934). *Early forerunners of man*. Baltimore: William Wood.

Crespi, L. P. (1942). Quantitative variation of incentive and performance in the white rat. *American Journal of Psychology*, **55**, 467-517.

Cross, H. A., & Harlow, H. F. (1965). Prolonged and progressive effects of partial isolation on the behavior of macaque monkeys. *Journal of Experimental Research in Personality*, **1**, 39–49.

Crozier, W. J. (1929). The study of living organisms. In C. Murchison (Ed.), *The foundations of experimental psychology*. Worcester, Mass.: Clark University Press.

Davis, R. T., Settlage, P. H., & Harlow, H. F. (1950). Performance of normal and brain-operated monkeys on mechanical puzzles with and without food incentive. *Journal of Genetic Psychology*, **77**, 305–311.

Dobzhansky, T. (1955). *Evolution, genetics, and man*. New York: Wiley.

Ellis, N. R., & Sloan, W. (1959). Oddity learning as a function of mental age. *Journal of Comparative & Physiological Psychology*, **52**, 228–230.

Erikson, E. H. (1950). *Childhood and society*. New York: Norton.

Fleure, H. J., & Walton, C. (1907). Notes on the habits of some sea anemones. *Zool. Anz.*, **31**, 212–220.

Flint, B. M. (1978). *New hope for deprived children*. Toronto: University of Toronto Press.

Flynn, J. P., & Jerome, E. A. (1952). Learning in an automatic multiple-choice box with light as incentive. *Journal of Comparative & Physiological Psychology*, **45**, 336–340.

French, J. W. (1940). Trial and error learning in paramecium. *Journal of Experimental Psychology*, **26**, 609–613.

Frisch, K. von. (1914). Der farbensinn und formensinn der biene. *Zool. Jahrb., Zool. Physiol.*, **35**, 1–18a.

Fuller, J. F., Easler, C. A., & Banks, E. M. (1950). Formation of conditioned avoidance responses in young puppies. *American Journal of Physiology*, **160**, 462–466.

Furchner, C. S., & Harlow, H. F. (1969). Preference for various surrogate surfaces among infant rhesus monkeys. *Psychonomic Science*, **17**, 279–280.

Gardner, B. T., & Gardner, R. A. (1971). Two-way communication with an infant chimpanzee. In A. M. Schrier & F. Stollnitz (Eds.), *Behavior of nonhuman primates* (Vol. 4). New York: Academic Press.

Gelber, B. (1952). Investigations of the behavior of *Paramecium aureles*. I. Modification of behavior after training with reinforcement. *Journal of Comparative & Physiological Psychology*, **45**, 58–65.

Ginsburg, N. (1957). Matching in pigeons. *Journal of Comparative & Physiological Psychology*, **30**, 261–263.

Goldstein, K. (1939). *The organism*. New York: American Book.

Goldstein, K., & Scheerer, M. (1941). Abstract and concrete behavior. An experimental study with special tests. *Psychology Monographs*, **53** (2, Whole No. 239).

Granit, R. (1955). *Receptors and sensory perception*. New Haven: Yale University Press.

Greenman, G. W. (1963). Visual behavior of newborn infants. In L. J. Stone, H. T. Smith, & L. B. Murphy (Eds.), *The conpetent infant*. New York: Basic Books.

Grether, W. F. (1939). Color vision and color blindness in monkeys. *Comparative Psychology Monographs, 15* (4, Whole No. 76).

Grether, W. F. (1940). A comparison of human and chimpanzee spectral hue discrimination curves. *Journal of Experimental Psychology, 26,* 394–403.

Groos, K. (1901). *The play of man.* New York: D. Appleton-Century.

Hamilton, G. V. (1911). A study in trial and error reactions in animals. *Journal of Animal Behavior, 1,* 33–66.

Hamilton, G. V. (1916). A study of perseverance reactions in primates and rodents. *Behavior Monographs, 3,* No. 2, 1–63.

Hamilton, W. F., & Coleman, T. B. (1933). Trichromatic vision in the pigeon as illustrated by the spectral hue discrimination curve. *Journal of Comparative Psychology, 15,* 183–191.

Haney, G. W. (1931). The effect of familiarity on maze performance of albino rats. *University of California Publications in Psychology, 4,* 319–333.

*Hansen, E. W. (1966). The development of maternal and infant behavior in the rhesus monkey. *Behavior, 27,* 107–149.

Harlow, H. F. (1939). Recovery of pattern discrimination in monkeys following unilateral osccipital lobectomy. *Journal of Comparative Psychology, 27,* 467–489.

Harlow, H. F. (1942). Responses by rhesus monkeys to stimuli having multiple sign values. In Q. McNemar & M. A. Merrill (Eds.), *Studies in personality.* New York: McGraw-Hill.

Harlow, H. F. (1943). Solution by rhesus monkeys of a problem involving the Weigl principle using the matching-from-sample method. *Journal of Comparative Psychology, 36,* 217–227.

*Harlow, H. F. (1945). Studies in discrimination learning by monkeys: Factors influencing the facility of solution of discrimination problems by rhesus monkeys. *Journal of Genetic Psychology, 32,* 213–227.

Harlow, H. F. (1949). The formation of learning sets. *Psychological Review, 56,* 51–65.

Harlow, H. F. (1950). Learning and satiation of response in intrinsically motivated complex puzzle performance by monkeys. *Journal of Comparative and Physiological Psychology, 43,* 289–294.

Harlow, H. F. (1951a). Primate learning. In C. P. Stone (Ed.), *Comparative psychology.* New York: Prentice-Hall.

Harlow, H. F. (1951b). Thinking. In H. Helson (Ed.), *Theoretical foundations of psychology.* New York: Van Nostrand.

Harlow, H. F. (1953). Mice, monkeys, men, and monkeys. *Psychological Review, 60,* 23–32.

Harlow, H. F. (1958). The evolution of learning. In A. Roe & G. G. Simpson (Eds.), *Behavior and evolution.* New Haven: Yale University Press.

Harlow, H. F. (1958). The nature of love. *American Psychologist, 13,* 673–685.

Harlow, H. F. (1959). The development of learning in the rhesus monkey. *American Scientist, 47,* 459–479.

*Harlow, H. F. (1962). The heterosexual affectional system in monkeys. *American Psychologist, 17,* 149.

Harlow, H. F. (1965). Sexual behavior in the rhesus monkey. In F. A. Beach (Ed.), *Sex and behavior.* New York: Wiley.

*Harlow, H. F. (1969). Age-mage or peer affectional system. In D. S. Lehrman, R. A. Hinde, & E. Shaw (Eds.), *Advances in the study of behavior* (Vol. 2). New York: Academic Press.

Harlow, H. F. (1971). The affectional systems. In H. F. Harlow, J. L. McGaugh, & R. F. Thompson (Eds.), *Psychology.* San Francisco: Albion Press.

Harlow, H. F., Davis, R. T., Settlage, P. H., & Meyer, D. R. (1952). Analysis of frontal and posterior association syndromes in brain-damaged monkeys. *Journal of Comparative and Physiological Psychology,* **45**, 410–429.

Harlow, H. F., Dodsworth, R. O., & Harlow, M. K. (1965). Total social isolation in monkeys. *Proceeding of the National Academy of Sciences,* **54**, 90–97.

Harlow, H. F., & Harlow, M. K. (1965). The affectional systems. In A. M. Schrier, H. F. Harlow, & F. Stollnitz (Eds.), *Behavior of nonhuman primates* (Vol. 2). New York: Academic Press.

*Harlow, H. F., Harlow, M. K., Dodsworth, R. O., & Arling, G. L. (1966). Maternal behavior of rhesus monkeys deprived of mothering and peer association in infancy. *Proceedings of the American Philosophical Society,* **110**, 58–66.

*Harlow, H. F., Harlow, M. K., & Hansen, E. W. (1963). The maternal affectional system of rhesus monkeys. In H. L. Rheingold (Ed.), *Maternal behavior in mammals.* New York: Wiley.

Harlow, H. F., & Hicks, L. H. (1957). Discrimination learning theory: Uniprocess vs. duoprocess. *Psychological Review,* **64**, 104–109.

*Harlow, H. F., Harlow, M. K., & Meyer, D. R. (1950). Learning motivated by a manipulation drive. *Journal of Experimental Psychology,* **40**, 228–234.

Harlow, H. F., Harlow, M. K., & Suomi, S. J. (1971). From thought to therapy: Lessons from a primate laboratory. *American Scientist,* **59**, 538–549.

Harlow, H. F., & Mears, C. E. (1978). Complex unlearned behaviors. In M. Lewis & L. Rosenblum (Eds.), *The development of affect: Genesis of behavior* (Vol. 1). New York: Plenum.

Harlow, H. F., & Novak, M. A. (1973). Psychopathological perspectives. *Perspectives in Biology & Medicine,* **16**, 461–478.

Harlow, H. F., & Rosenblum, L. A. (1971). Maturational variables influencing sexual posturing in infant monkeys. *Archives of Sexual Behavior,* **1**, 73–78.

Harlow, H. F., Schiltz, K. A., Blomquist, A. J., & Thompson, C. I. (1970). Effects of combined frontal and temporal lesions on learned behaviors in rhesus monkeys. *Proceeding of the National Academy of Sciences,* **66**, 577–582.

Harlow, H. F., & Spaet, T. (1943). Problem solution of monkeys following bilateral removal of the prefrontal areas. *Journal of Experimental Psychology,* **33**, 500–507.

Harlow, H. F., & Suomi, S. J. (1970). The nature of love—simplified. *American Psychologist,* **25**, 161–168.

Harlow, H. F., & Suomi, S. J. (1971a). Production of depressive behaviors in young monkeys. *Journal of Autism and Childhood Schizophrenia,* **1**, 246–255.

302 HARRY F. HARLOW AND CLARA MEARS

Harlow, H. F., & Suomi, S. J. (1971b). Social recovery by isolation-reared monkeys. *Proceedings of the National Academy of Sciences,* **68**, 1534–1538.
Harlow, H. F., & Yudin, H. (1933). Social behavior of primates. I. Social facilitation of feeding in the monkey and its relation to attitudes of ascendance and submission. *Journal of Comparative Psychology,* **16**, 171–185.
Harlow, H. F., & Zimmermann, R. R. (1958). The development of affectional responses in infant monkeys. *Proceedings of the American Philosophical Society,* **102**, 501–509.
Harlow, H. F., & Zimmermann, R. R. (1959). Affectional responses in the infant monkey. *Science,* **130**, 421–432.
Harlow, M. K., Harlow, H. F., Eisele, C. D., & Ruppenthal, G. C. *Rhesus macaque paternal behavior in a nuclear family situation.* Cited in Suomi et al., 1973. Cited in Lamb, M. E., 1976.
Harper, L. V., & Sanders, K. M. (1975). Preschool children's use of space. Sex differences in outdoor play. *Developmental Psychology,* **2**, 119.
Harris, J. D. (1943). The auditory acuity of pre-adolescent monkeys. *Journal of Comparative Psychology,* **35**, 255–265.
Hayes, C. (1951). *The ape in our house.* New York: Harper.
Head, H. (1926). *Aphasia and kindred disorders of speech.* London: Cambridge University Press.
Hebb, D. O. (1949). *The organization of behavior.* New York: Wiley.
Hediger, H. (1955). *Studies of the psychology and behavior of captive animals in zoos and circuses.* New York: Criterion Books.
Helfer, R. E., & Kempe, C. H. (1968). *The battered child.* Chicago: University of Chicago Press.
Hicks, L. H. (1956). An analysis of number-concept formation in the rhesus monkey. *Journal of Comparative & Physiological Psychology,* **49**, 212–218.
Hinde, R. A. (1974). *Biological bases of human social behaviour.* London: McGraw-Hill, 149–153.
Hinde, R. A. (1975). Mothers and infants roles: Distinguishing the questions to be asked. In *Parent-Infant Interactions Ciba Founation Symposium,* **33**. Amsterdam: Elsevier Excerpta Medica.
Hinde, R. A., Spencer-Booth, Y., & Bruce, M. (1966). Effects of 6-day maternal deprivation on rhesus monkey infants. *Nature,* **210**, 1021–1033.
Hodos, W. (1970). Evolutionary interpretation of neural and behavioral studies of living vertebrates. In F. O. Schmidt (Ed.), *The neurosciences: Second study program.* New York; Rockefeller University Press.
Horenstein, B. (1951). Performance of conditioned responses as a function of strength of hunger drive. *Journal of Comparative and Physiological Psychology,* **44**, 210–224.
Hovey, H. B. (1929). Associative hysteresis in flatworms. *Physiol. Zool.,* **2**, 322–333.
Hull, C. L. (1952). *A behavior system.* New Haven: Yale University Press.
Hunter, W. F. (1913). The delayed reaction in animals and children. *Behavior Monographs,* **2**, 21–30.

Hurlock, E. B. (1934). Experimental investigations of childhood play. *Psychological Bulletin*, 31, 47–66.

Itani, J. (1959). Paternal care in the wild Japanese monkey, *Macaca fuscata*. *Primates*, 2, 61–93.

Jacobsen, C. F. (1935). An experimental analysis of the frontal assocation areas in primates. *Archives of Nervous and Mental Diseases*, 82, 1–14.

Jacobsen, C. F. (1936). Studies of cerebral function in primates. *Comparative Psychology Monographs*, 13, 3–60.

Jennings, H. S. (1906). *Behavior of the lower organisms*. New York: Columbia University Press.

Jensen, K. (1932). Differential reactions to taste and temperature stimuli in newborn infants. *Genetic Psychology Monographs*, 12, 361–476.

Johnson, M. W. (1971). In R. E. Herron & B. Sutton-Smith (Eds.), *Child's play*. New York: Wiley.

Kaufman, I. C., & Rosenblum, L. A. (1967). The reaction to separation in infant monkeys: Anaclitic depression and conservation-withdrawal. *Psychosomatic Medicine*, 29, 648–675.

Keller, F. S. (1941). Light-aversion in the white rat. *Psychological Record*, 4, 235–250.

Kinsey, A. C., Pomeroy, W. B., & Martin, C. E. (1948). *Sexual behavior in the human male*. Philadelphia: W. B. Saunders.

Klüver, H. (1933). *Behavior mechanisms in monkeys*. Chicago: University of Chicago Press.

Klüver, H. (1935). An auto-multi-stimulation reaction board for use with sub-human primates. *Journal of Psychology*, 1, 123–127.

Koch, S., & Daniel, W. J. (1945). The effect of satiation on the behavior mediated by a habit of maximum strength. *Journal of Experimental Psychology*, 35, 162–185.

Köhler, W. (1925). *The mentality of apes*. New York: Harcourt & Brace.

Kohts, N. (1928). Recherches sin l'intelligence du chimpanze par la methode de choix d'apres mdele.

Korner, A. F. (1974). The effect of the infant's state, level of arousal, sex, and antigenetic stage on the care giver. In M. Lewis & L. Rosenblum (Eds.), *The effect of the infant on its caregivers*. New York: Wiley.

Krechevsky, I. (1932). "Hypothesis" vs. "chance" in the pre-solution period in sensory discrimination learning. *University of California Publications in Psychology*, 6, 27–44.

Kubie, L. S. (1953). In M. Heiman (Ed.), *Psychoanalysis and social work*. New York: International Universities Press.

Kulka, A., Fry, C., & Goldstein, F. J. (1960). Kinesthetic needs in infancy. *American Journal of Orthopsychiatry*, 30, 362–571.

Kummer, H. (1968). Two variations in the social organizations of baboons. In P. C. Jay (Ed.), *Primates*. New York: Holt, Rinehart & Winston.

Lamb, M. E. (1975). Fathers: Forgotten contributors to child development. *Human Development*, 18, 245–266.

Lamb, M. E. (1976). *The role of the father in child development*. New York: Wiley.

Lamb, M. E. (1978). *Social and personality development.* New York: Holt, Rinehart, & Winston.

Langmeier, Jr., & Matějček, Z. (1976). *Psychological deprivation in childhood* (G. L. Mangan, Ed.). New York: Wiley.

Langworthy, O. R. (1928). The behavior of pouch-young opossums correlated with the myelinization of tracts in the nervous system. *Journal of Comparative Neurology*, 46, 201–248.

Lashley, K. S. (1929). *Brain mechanisms and intelligence: A quantitative study of injuries to the brain.* Chicago: University of Chicago Press.

Lee, L. C. (1973). *Social encounters of infants: The beginnings of popularity.* Paper presented at the International Society for the Study of Behavioral Development, Ann Arbor, Michigan.

Lennox, M. A. (1956). Geniculate and cortical responses to colored light flash in the cat. *Journal of Neurophysiology*, 19, 271–279.

Lewis, M. (1976). *The origins of self competence.* Paper presented at NIMH Conference on Mood Development, Washington, D.C.

Lewis, M., & Brooks, J. (1978). Self-knowledge and emotional development. In M. Lewis & L. Rosenblum (Eds.), *The development of affect. Genesis of behavior* (Vol. 1). New York: Plenum.

Loeb, J. (1918). *Forced movements, tropisms and animal conduct.* Philadelphia: Lippincott.

Maier, N. R. F., & Schneirla, T. C. (1935). *Principles of animal psychology.* New York: McGraw-Hill.

Martin, A. S., & Tyrrell, D. J. (1971). Oddity learning following object-discrimination learning in mentally retarded children. *American Journal of Mental Deficiency*, 75, 504–509.

Maslow, A. H. (1943). A theory of human motivation. *Psychological Review*, 50, 370–396.

Mason, W. A., & Sponholz, R. R. (1963). Behavior of rhesus monkeys raised in isolation. *Psychiatric Research*, 1, 1–8.

Mast, S. O., & Pusch, L. C. (1924). Modification of response in amoeba. *Biological Bulletin*, 46, 55–60.

McCulloch, T. L., & Haselrud, G. M. (1939). Affective responses of an infant chimpanzee reared in isolation from its kind. *Journal of Comparative Psychology*, 28, 437–445.

McKinney, W. T., & Bunney, W. E. (1969). *Archives of General Psychiatry*, 21, 240.

McKinney, W. J., Jr., Eising, R. G., Moran, E. C., Suomi, S. J., & Harlow, H. F. (1971). Effects of reserpine on the social behavior of rhesus monkeys. *Diseases of the Nervous System*, 32, 735–741.

Mears, C. E. (1978). Play and development of cosmic confidence. *Developmental Psychology*, 14, 371–378.

Mears, C. E., & Harlow, H. F. (1975). Play, early, and eternal. *Proceedings of the National Academy of Sciences*, 72, 1878.

Meyer, D. R. (1951). Food deprivation and discrimination reversal learning by monkeys. *Journal of Experimental Psychology*, 41, 10–16.

Miller, N. E. (1951). Learnable drives and rewards. In S. S. Stevens (Ed.), *Handbook of experimental psychology*. New York: Wiley.

Miller, R. E., Murphy, J. V., Mirsky, I. A. (1959). Relevance of facial expression and posture as cues in communication of affection between monkeys. *Archives of General Psychiatry, 1*, 480–488.

Mitchell, G. D., & Clark, D. L. (1968). Long-term effects of social isolation in nonsocially adapted rhesus monkeys. *Journal of Genetic Psychology, 113*, 117–128.

Moon, L. E., & Harlow, H. F. (1955). Analysis of oddity learning by rhesus monkeys. *Journal of Comparative & Physiological Psychology, 48*, 188–194.

Nissen, H. W. (1931). A field study of the chimpanzee. *Comparative Psychology Monographs, 8* (1, Whole No. 36).

Nissen, H. W. (1951). Analysis of conditioned reaction in chimpanzees. *Journal of Comparative and Physiological Psychology, 44*, 9–16.

Nissen, H. W., Blum, J. S., & Blum, R. A. (1948). Analysis of matching behavior in chimpanzee. *Journal of Comparative & Physiological Psychology, 41*, 62–74.

Novak, M. A. (1974). *Fear attachment relationships in infant and juvenile rhesus monkeys*. Unpublished doctoral dissertation, University of Wisconsin, Madison, #74-00490.

Novak, M. A., & Harlow, H. F. (1975). Social recovery of monkeys isolated for the first year of life. 1. Rehabilitation and therapy. *Developmental Psychology, 11*, 453–465.

Passman, R. H., & Weisbey, P. (1975). Mother and blankets as agents for promoting play and exploration by young children in a novel environment. The effects of social and nonsocial attachment objects. *Developmental Psychology, 89*, 285–292.

Pavlov, I. P. (1927). *Conditioned reflexes* (G. V. Anrep, trans.). London: Oxford University Press.

Pratt, C. L. (1969). *The developmental consequences of variations in early social stimulation*. Unpublished doctoral dissertation, University of Wisconsin, #70-03665.

Premack, D. (1971). On the assessment of language competence in the chimpanzee. In A. M. Schrier & F. Stollnitz (Eds.), *Behavior of nonhuman primates* (Vol. 4). New York: Academic Press.

Pribram, K. H. (1955). Toward a science of neuropsychology (method and data). In R. A. Patton (Ed.), *Current trends in psychology*. Pittsburgh: University of Pittsburgh Press.

Provence, S., & Lipton, R. C. (1962). *Infants in institutions*. New York: International Universities Press.

Reeves, C. (1919). Discrimination of light of different wave-lengths by fish. *Comparative Psychology Monographs, 4*, 57–83.

Richter, C. P. (1927). Animal behavior and internal drives. *Quarterly Review of Biology, 2*, 307–343.

Richter, C. P. (1942-1943). Self regulatory functions in animals and human beings. *Harvey Lectures Series, 38*, 63–103.

Roberts, K. E. (1933). Learning in pre-school and orphanage children: An experimental study of ability to solve different situations according to the same plan. *University of Iowa Studies in Child Welfare,* **7** (3, Whole No. 251).

Robinson, J. S. (1953). Stimulus substitution and response learning in the earthworm. *Journal of Comparative & Physiological Psychology,* **46,** 262–266.

Rosenblum, L. A., & Harlow, H. F. (1963). Approach-avoidance conflict in the mother-surrogate situation. *Psychological Reports,* **12,** 83–85.

Rosenblum, L. A., & Kaufman, I. C. (1968). Variations in infant development and response to maternal loss in monkeys. *American Journal of Orthopsychiatry,* **38,** 418–426.

Rosevear, J. A. Y. (1970). *Early sex differentiation and diurnal patterns of behavior in the rhesus monkey.* Unpublished Master's thesis, University of Wisconsin.

Rowland, G. L. (1964). *The effects of total social isolation upon learning and social behavior in rhesus monkeys.* Unpublished doctoral dissertation, University of Wisconsin.

Sackett, G. P. (1966). Monkeys reared in visual isolation with pictures as visual input. *Science,* **154,** 1468–1472.

Sackett, G. P. (1968). The persistence of abnormal behaviour in monkeys following isolation rearing. In R. Porter (Ed.), *The role of learning in psychotherapy.* London: J. & A. Churchill Ltd.

Schiller, P. (1949). Delayed detour response in the octopus. *Science,* **42,** 262–269.

Schmale, A. (1971). *The role of depression in health and disease.* Paper presented at 137th annual convention, AAAS, Chicago, Illinois, December, 1971.

Schneirla, T. C. (1934). The process and mechanism of ant learning. *Journal of Comparative Psychology,* **17,** 303–329.

*Seay, B., Hansen, E., & Harlow, H. F. (1962). Mother-infant separation in monkeys. *Journal of Child Psychology & Psychiatry,* **3,** 123–132.

Seay, B., & Harlow, H. F. (1965). Maternal separation in the rhesus monkey. *Journal of Nervous and Mental Diseases,* **140,** 434–441.

Seward, J. P., Levy, N., & Handlon, J. H., Jr. (1950). Incidental learning in the rat. *Journal of Comparative & Physiological Psychology,* **43,** 240–251.

Sheffield, F. D., & Roby, T. B. (1950). Reward value of a non-nutrient sweet taste. *Journal of Comparative & Physiological Psychology,* **43,** 471–481.

Singh, J. A. L., & Zingg, R. M. (1966). *Wolf-children and feral man.* U.S.A.: Archon Books.

Singh, S. D. (1969). Urban monkeys. *Scientific American,* **24,** 108–115.

Spaet, T., & Harlow, H. F. (1943). Solution by rhesus monkeys of multiple sign problems utilizing the oddity technique. *Journal of Comparative Psychology,* **35,** 119–132.

Spence, K. W. (1950). Cognitive versus stimulus-response theories of learning. *Psychological Review,* **57,** 159–172.

Spencer, H. (1873). *The principles of psychology.* New York: D. Appleton-Century.

Spitz, R. A. (1946). Hospitalism: A followup report. *Psychoanalytic Study of the Child,* 2.

Stone, C. P. (1929). The age factor in animal learning: II. Rats on a multiple light discrimination box and a difficult image. *Genetic Psychology Monographs,* 6, No. 2, 125–202.

Stone, L. J., Smith, H. T., & Murphy, L. B. (Eds.) (1973). *The competent infant.* New York: Basic Books.

Strassburger, R. C. (1950). Resistance to extinction of a conditioned operant as related to drive level at reinforcement. *Journal of Experimental Psychology,* 40, 473–487.

Suomi, S. J. (1973). Surrogate rehabilitation of monkeys reared in total social isolation. *Journal of Child Psychology & Psychiatry,* 14, 71–77.

Suomi, S. J., Collins, M. L., & Harlow, H. F. (1973). Effects of permanent separation from mother on infant monkeys. *Developmental Psychology,* 9, 376–384.

Suomi, S. J., & Harlow, H. F. (1969). Apparatus conceptualization for psychopathological research in monkeys. *Behavior Research Methods and Instrumentation,* 1, 247–250.

Suomi, S. J., & Harlow, H. F. (1971). Monkeys at play. *Natural History,* 80, 72–76.

Suomi, S. J., Harlow, H. F., & Domek, C. J. (1970). Effect of repetitive infant-infant separation of young monkeys. *Journal of Abnormal Psychology,* 76, 161–172.

Suomi, S. J., Harlow, H. F., & Kimball, S. D. (1971). Behavioral effects of prolonged partial social isolation in the rhesus monkey. *Psychological Reports,* 29, 1171–1177.

Tanner, J. M. (1970). Physical growth. In P. H. Mussen (Ed.), *Carmichael's manual of child psychology* (Vol. 1). New York: Wiley.

Teuber, H. L. (1955). Physiological psychology. *Annual Review of Psychology,* 6, 267–296.

Thorndike, E. L. (1911). *Animal intelligence.* New York: Macmillan.

Thorpe, W. H. (1943). Types of learning in insects and other arthropods. *British Journal of Psychology,* 33, 220–234.

van Wagenen, G. (1950). The monkey. In E. J. Farris (Ed.), *The care and breeding of laboratory animals.* New York: Wiley.

Warren, J. M. Personal communication.

Watson, J. B. (1914). *Behavior: An introduction to comparative psychology.* New York: Holt.

Watson, J. B. (1924). *Behaviorism.* New York: Norton.

Weigl, E. (1941). On the psychology of so-called processes of abstraction. *Journal of Abnormal & Social Psychology,* 36, 5–33.

Weinstein, B. (1955). The evolution of intelligent behavior in rhesus monkeys. *Genetic Psychology Monographs,* 31, 3–48.

Winsten, B. (1945). The evolution of intelligent behavior in rhesus monkeys. *Genetic Psychology Monographs,* 31, 3–48.

Wodinsky, J., & Bitterman, M. E. (1953). The solution of oddity-problems by the rat. *American Journal of Psychology,* 64, 137–140.

Wolpe, J. (1958). *Psychotherapy by reciprocal inhibition.* Stanford: Stanford University Press.

Yerkes, R. M. (1912). The intelligence of earthworms. *Journal of Animal Behavior,* **2**, 332–338.

Yerkes, R. M., & Dodson, J. D. (1908). The relation of strength of stimulus to rapidity of habit formation. *Journal of Comparative Neurology & Psychology,* **18**, 459–482.

Young, M. L., & Harlow, H. F. (1943). Solution by rhesus monkeys of a problem involving the Weigl principle using the oddity method. *Journal of Comparative Psychology,* **35**, 205–217.

Young, P. T. (1949). Food-seeking drive, affective process, and learning. *Psychological Review,* **56**, 98–121.

Zeaman, D. (1949). Response latency as a function of amount of reinforcement. *Journal of Experimental Psychology,* **39**, 466–483.

Zeaman, D., & House, B. J. (1950). Response latency at zero drive after varying number of reinforcements. *Journal of Experimental Psychology,* **40**, 570–583.

INDEX

Affectional systems (*see also* Infant, Maternal, Paternal, Peer, Heterosexual)
derived drives, 102
emotional sequencing and maturation, 164-165, 281
five love systems, 161-164
psychoanalytic theories, 103
Age, relationship between man and monkey, 3-4
Agemate love (*see* Peer love)
Agemate separation (*see* Separation)
Aggression, 164-165, 197-200, 210-213, 281-282, 288
amelioration of, 164-165, 198-200, 213, 281-282
late maturation, 197-198, 212-213
sex differences in threat and aggression, 210-213
Apparatus
learning: Wisconsin General Test Apparatus, 14-15

motivation and love: Butler box, baby and adult, 83, 116
Attachment, avian, ungulate, and primate, 174-175
Autistic child, 169

Bodily contact acceptance, 166, 193, 261-262
Breeding colony, 4
Breland phenomenon, 17

Caretaker (*see* Maternal and Paternal)
Clinging, 174-175
Communication, 166-169, 178-180, 280
Contact comfort, 104-110, 133-134, 144, 274-275
Cortical localization, 6-7, 274
equipotentiality, Lashley, specificity, Jacobsen, 77
Curiosity (*see* Motivation, Exploration)

Depression (*see also* Separation)
  adolescent, 238-239
  anaclitic, 225-237
Deprivation of loves (*see also* Separation, Isolation)
  agemate, 197
  mother, 171, 197, 261-262, 289
  timing effects, 170
Developmental comparison of man and monkey, 2, 3, 7-9
Developmental sequencing of individual behaviors: bodily contact, communication, exploration, security, peragration, 148, 166-170
Drives vs. incentives, 7 (*see also* Motivation)

Emotional sequencing, 164-170, 177
Environment, feral vs. optimal, 5-6
  effects on learning of deprived and enriched, 283-284
Exploration, 83-86, 96-99, 146, 150-152

Father (*see* Paternal)
Fear, 111-115, 164-165, 175-179, 234, 288
Feedback, interaction, 168-169, 182
Following, 174-175

Generalization of behavior, 1-2, 4, 189
Gestural communication, 178-180
Grooming, 209

Heredity and environment, 201
Heterosexual love, 162, 164, 166
  antecedent loves, importance of, successful sex or love, 213
  likes and dislikes, 204-205
    Tommy the Terrible, 203-204
  sex characteristics, overlapping, 206
  sex differences, subhuman primates and some human primates, 195-197, 206-213, 282
  sex preferences, 202-204

Imitation, 180
Imprinting or following, 124, 174-175
Incentives, 7, 78-79, 82, 97-99
Individual play, 151-153
Infant independence, 180, 183, 191, 283
Infant-mother love, 162
  affectional variables, 104-114

contact comfort, 104-105, 108-110, 144, 274-275
  facial, 130, 132-133
  lactation, 106-108, 129-130, 278
  rocking motion, 133-134, 148, 278
  security and trust, 110-116, 127, 169-170, 178, 276-277
  temperature, 133, 135-138, 279
  persistence and retention, 117-124, 230
Infant-mother separation (*see* Separation)
Infant preferences, parental and agemate, 204-206
Innate responses, simple and complex, 201-202
Institutional care: hospitals, orphanages, foster homes, 166-167, 227
Isolation, partial, 219, 241-244, 252
  age changes, 254
  compulsive behavior syndrome, 244, 252
  depression and loss of social behaviors, play and sex, 252, 285, 288
  schizoid type postures, 254, 286
Isolation, total social, definition and examples, 241-244, 287
  age effects: 3 months, 244-248
    6 months, 246-248, 257-258, 287-288
    12 months, 245, 251-252, 258, 288
  depression and loss of social behaviors, play, sex, 246-251
  infantile, maladaptive behaviors, 247
  maintenance of health, 249
  prolonged effects, 249, 252

Latency period, 197, 213
Learning
  adaptation, 30-31
  age changes through maturation and learning, 21-25, 31, 33-36, 39-41
  ambiguity of reward, 24
  apparatus, 14-15
  complex and simple learned responses, 36, 71-72, 201-202
  effects of specific environments, 283-284
  excitation and inhibition, 54-59
  insight learning, 66-67
  internal drive theory, effect on learning, 90-95
  symbolic, 72
  tool use, 69-71
Learning, evolution of

by geologic periods, sea and land dwelling animals, 44, 45
from anatomical point of view, 45-46
receptor development, visual mechanisms, 46-50
through problems of increasing complexity, 51-53
Learning sets, nature of, 35-36, 62-63, 66, 71
applications, 72-73
cortical localization, 7
Hull's theory, 69
reversal, 67-69
trials
importance of *Trial 2*, 35, 64-67
number of, 64-66
trial and errors, 63
Learning tests
delayed response and individual differences, 31-35
discrimination, spatial and object, 17-18, 21, 23, 30-31, 51, 56
Hamilton perseverance, 37-39
matching and oddity, 19-22, 39-40, 284
multiple sign problems, 19-23, 39-40, 280
Weigl, 22-23, 51-53
Love (*see also* Emotional sequencing, Affectional systems), successful love, 213

Maternal love, 162 (*see also* Infant-mother, Surrogate)
communication and interaction, 166-169
dangers and strangers, 176, 177, 183
deprivation, 182-183
live mother or caretaker vs. surrogate, 279-280
communication, 178-180
fear, 175-179
infant independence, 183
play or social roles, 142-144, 180-183
protection, 176-177, 183, 210
maternal love retention, 117-124
second and third children, 282-283
Maturation
aggression, 211-212; emotional amelioration of, 213
fear, 163-165, 234
learning, 31, 33-36
monkey vs. human being, 4
play, 264-295

pleasant vs. unpleasant emotions, 164-165
self-motion play, 157
Monkeys as research subjects, 1-5, 8, 9, 218, 225
learning, 13, 27-29, 41
limitations of human subjects, 28, 29, 243
love, 103
psychopathology, 218, 222, 225
Motherless mothers, 261-262, 289-290
Motivation: internal and external motives, 78-95
external motives, incentives, 78-80, 82-86, 89-90, 96-99
complex, unlearned behaviors: curiosity (Butler box), exploration, 80, 83-86, 96-90
effect of brain damage, 86
ontogenetic approach, 89-90
theories: pain and pleasure, Thorndike, 90; tropistic, Loeb, 89
reinforcement, 85-86
variation of strength, 89-90
internal, visceral drives, 78-79, 81-82, 87-98, 102, 256
application to simple learning and behavior, 94-95; brevity, 92
derived drives, Hull and Spence, 79, 81-82, 92
homeostasis, Richter, 79
hunger vs. happiness and learning, 78, 92-94, 97-98
hypothalamus, Cannon, 78-79
inhibition of learning, deprivation studies, 93-94
instinct or tissue tension, Watson, behaviorism, 90
love or affection, 102

Nuclear family, 184-185, 238, 282, 283

Passivity, 209-210
Paternal love, 162-164
aggression, paternal role in amelioration, 164, 182, 185
as caretakers, 187
communication, 168
deprivation of, 187
play, 166, 175, 186-187, 257-283
protection, 163-164, 184-185, 282
Peer love, 162, 189-200, 191-192

aggression amelioration, 199
appearance, 191-192, 194
breadth, depth and extent, 189, 191-192
latency period, 197, 213
loss or lack of playmates, temporary, 192;
    prolonged, 197; of mothers, 191, 197
play as developer, 142-145, 191-195
social and sexual roles, 193-199
Play
apparatus and play, 150-151, 154-155
development of self-confidence and com-
    petence, 155-157, 170
developmental picture, 143-144, 150-153
emotional security, 127
exploration and play, 146, 150-152
individual play, 151-153
mastery mechanism, 155-156, 170
peragration or self-motion play, 146-157,
    170, 191-192
sex differences in play, 195-197, 208
sexual roles and play, 144-145, 196
social play, 149, 151:
    approach-avoidance, 144, 193-195;
    rough-and-tumble, 144, 193-195, 208-
    209

Rearing conditions, 5, 204
Rehabilitation of psychopathic monkeys,
    257-271
early techniques, 257
    social agemates, 259-260
partial improvement: motherless mother,
    261-262, 289-290; surrogates, 261-262
rehabilitation, social criteria for, 267
    rehabilitation of 6-month isolates,
    263-268
    rehabilitation of 12-month isolates,
    268-271
Rocking motion, 133-134, 148, 278

Security and trust, 110-115, 127, 169-170,
    178, 276, 277
Separation studies, depression

agemate repetitive separations, 234-237
persistence of infantile behaviors, 235
infant-mother, 225-237; hospitalized or
    orphaned infants, 227
    age-sex effects, 233
    attachment patterns prior to separation,
    233
    depression of play and social behaviors,
    232, 235, 289
    psychological vs. physical separation,
    226
    researches of Spitz and Bowlby, 225-
    227
    stages of protest, despair, and reunion,
    227-228, 230, 231, 289-290
    war-separated children, 226, 232
Sex differences, 195-197, 206-213, 282
Silly grin, 178
Stimulus facilitation, interference, generali-
    zation, 23-26
Surrogate mother, cloth and wire, 106-109,
    111-127, 274-275
covers, 133-134
rejecting, 285
simplified surrogate, 133-134
Symbiosis, 183

Therapy, psychopathological
infant as inadvertent therapist, 261-262
infant as planned therapist, 263-271
    6-month isolates, 3-month junior thera-
    pists, 263-268
    12-month isolates, 268-271; age of
    therapists, 269
play as therapy, 263-268, 270
procedure of gradual introduction of
    change, 263-264, 269, 271
self-paced therapy, 269, 270
social contact acceptance: motherless
    mothers, 261-262, 289-290; surrogate,
    261-262
Threat, sex differences, 211

DATE DUE
